Michael Gage *Maritz Vandenberg*

Authors

MICHAEL GAGE, BArch (Hons) RIBA, AIAA is
senior lecturer at the Polytechnic of Central London.
He was formerly Editor of *Precast Concrete*, **and had**
responsibility for courses for architects and students
at the C&CA Training and Conference Centre;
thereafter he was head of information for the British
Ready Mixed Concrete Association, and publications
manager for Ready Mixed Concrete Limited.

MARITZ VANDENBERG, BA (Hons) Arch, is
Technical Editor of *The Architects' Journal.*

HARD LANDSCAPE IN CONCRETE

by Michael Gage
and Maritz Vandenberg

The Architectural Press,
London

Halsted Press Division
John Wiley & Sons,
New York

Introduction

712
G121L
76-3053

This book provides designers with a comprehensive guide to the use of concrete in hard landscape. In the initial section, the authors discuss the general problem of urban design, with special reference to the use of hard surfaces in the formation of urban spaces. This is followed by a series of information sheets giving guidance on how to achieve desired specific finishes. The earlier sections draw on the ideas put forward by Gordon Cullen, Nan Fairbrother, Elizabeth Beazley and the other writers credited in the text (their publications are listed in the References section, page 166); whereas the information sheets draw on the research work and publications of the Cement and Concrete Association (C & CA).
The authors acknowledge, and are most grateful for, the considerable advice and assistance given by Nicolette Franck of the C&CA; Philip Gooding, formerly head of information, training and publishing, C&CA; Tom Kirkbride, technical director of the British Ready Mixed Concrete Association; and Dr Maurice Levitt of John Laing & Son Ltd.

British edition

ISBN 0 85139 277 6
First published in book form 1975
by The Architectural Press Ltd
© Michael Gage and Maritz Vandenberg 1975

Published in the USA by
Halsted Press, a Division of
John Wiley & Sons Inc,
New York

Library of Congress Cataloging in Publication Data

Gage, Michael Terence
  Hard Landscape in Concrete.

Bibliography p166
1. Concrete construction. 2. Landscape architecture.
I. Vandenberg, Maritz, joint author.
II. Title.
TA681.G23 712 75-31700

ISBN 0-470-28913-9

Printed in Great Britain by
Diemer & Reynolds Ltd, Bedford

Contents

1 The urban scene

1 *King's Lynn provides good example of townscape qualities often missing in modern cities, both in terms of space manipulation and detailing of surfaces.*

1.01 Introduction

Cities consist of buildings. But, equally important, they consist of the connective web of open space between those buildings, linking them and interpenetrating their interiors. The art of urban design lies in manipulating these external spaces so as to form a continuous yet varied urban landscape that is both functionally sensible and aesthetically satisfying. The complex networks of urban space found in medieval European towns and cities (and in many other traditional settlements all over the world) still give us immense aesthetic satisfaction—their tight spatial scale, richness of form and texture, and overall visual coherence create an environment felt by most observers to possess a 'humanity' sadly lacking in most modern city spaces, **1**.

This is not the place to examine how such influences as the needs of the motorcar, the aesthetic insensitivity of building codes and by-laws, the spatially disruptive effect of tower blocks, and the lack of a mature design vocabulary for new building materials and production techniques have coarsened, and in some cases destroyed, the traditional urban landscape. Nor does it fall within the scope of this book to offer advice on fundamental problems of urban form (ie, the large-scale organisation of urban spaces and elements). But it will nevertheless be useful to introduce the subject of hard landscape in concrete by looking briefly at the nature of urban space, so that the specific problems of site design, material and component selection, and the detailing of shapes and surfaces can be seen in the wider perspective.

2

3a

1.02 The nature of urban space

It is helpful to think of cities as consisting of two kinds of external space—paths and places.

Paths cater for movement, **2**. They enable us to get from where we are to where we want to be. They must not only facilitate physical movement (by vehicle or on foot), but they must also help us to orient ourselves, guide us and help us find our way. By 'paths' we mean the roads, pavements, alleys, lanes, footways, steps and ramps which form our routes through the urban web.

Places, on the other hand, are the nodes where movement comes to a stop. They are the parks, squares, courtyards, gardens and (at the smallest end of the scale) sitting areas where we can work, play, rest, or chat with friends, **3a, b**.

The division of urban space into paths and places is not, of course, clear-cut—the two functions are usually more ambiguous, and frequently overlap. But one function is usually primary and the other secondary, and this influences design. A street, mall or pavement, for example, is first a channel for movement and the designer cannot afford to ignore this fact: attention must be given to such aspects as traffic-bearing surfaces, gradients, traffic signs and signals. But the presence of stopping-places along the way (shop windows and shop entrances; news stands; benches in quiet corners; little parks) gives it also the character of a sequence of places, **4a**. And conversely, an urban square or courtyard is primarily a place, but it usually has a secondary system of pathways crossing or encircling it, **4b**.

2 *Example of 'paths' (both vehicle and pedestrian); the Hofgarten in Düsseldorf.*
3 *Two examples of 'places'* **a** *is large-scale and urban in character (Bath); and* **b** *small-scale and domestic (a town garden using exposed concrete paving, cobbles and concrete plant containers).*
4a *Example of urban space that is primarily a 'path', with a series of secondary 'places' strung along its length: the Weinstrasse in Munich, rebuilt as a pedestrian precinct with its irregular, quaint, crooked lines intact.* **b** *This shopping courtyard in Copenhagen conversely is primarily a 'place' traversed by a network of 'paths'. Interesting floorscape provided by precast concrete paving slabs and granite setts.*

4a

3b

4b

Moving through the city, the urban dweller therefore traverses a series of paths each of which has a sequence of places strung along its length like beads on a string. The aim of landscape design should be to realise the latent character of each path and place to the full, bringing out its unique possibilities, **5** and exploiting contrasts in function, scale and character. Ideally, a trip through the city should be (in Gordon Cullen's words*) a 'plastic experience'—a journey through a sequence of pressures and vacuums, constraints and reliefs, exposures and enclosures as the pedestrian or passenger moves from the constriction of the alley to the wideness of the square, from the containment of the street to the sudden revelation of the fly-over, in a constantly changing series of emergent views, **6**.

5 *Environmental potential of site exploited to the utmost. Stepping stones over water provide exciting, slightly hazardous access to safe place formed around tree.*

6 *The contrast between the scale and form of the pedestrian and vehicle access in Liverpool is heightened by the series of emergent views along the pedestrian route.*

5

6

1.03 Landscape design

Landscape design should always be firmly rooted in the two foundations of *existing local character*, and *proposed landscape function*.

Existing local character

Character of site

Each locality or site has its own unique existing qualities, and while these need not determine the design concept produced by landscaper or architect, they will certainly provide a valid and useful starting point for design. For instance, the designer should first approach any landscaping problem by checking the following points:

● What is the *land form* of the site—is it level, or steeply sloping; flat, or contoured; concave, or convex? Each of these formations suggests a different design concept.

● Does the site have any pronounced sense of *position* in relation to the surrounding or adjacent domain? Is it raised above, or sunk below, or set at an angle to, the dominant local ground level? Is it exposed or submerged; part of the larger domain or sharply differentiated from it by level or slope?

What are the *views and vistas* available to those using the

site—both users at rest (sitting on benches or relaxing on lawns) and users on the move (walking, climbing steps, driving)? In the case of people on the move, the concept of 'serial vision' is particularly important. New and perhaps unexpected views are constantly revealed as the pedestrian moves along, some of them emerging gradually, others revealed with dramatic suddenness; and this tension between the existing view and the emerging view is, as Gordon Cullen points out, a 'tool with which human imagination can begin to mould the city into a coherent drama'*.

It is on the basis of factors such as these that the site's landscape potential can be assessed. Some sort of *visual record* of the proposed site should be made before design is started, analogous to the physical record of soil and climatic data, land use and so on, that is commonly undertaken by architects and landscapers. Suggested methods of recording a visual analysis of this kind are described in the *AJ Handbook of urban landscape*† and summarised below.

At the very least, the spatial form of the site should be mentally absorbed by walking across it several times, and recorded by means of photographs and sketches. But where the site is large and complex it will be useful also to use the 'isovist' method of recording on a series of plans the edges of those objects which form boundaries to one's vision from selected positions in each separate space zone on the site; lines forming the boundaries of areas of 'dead ground'; and lines forming visual

*The concise townscape, by Gordon Cullen, page 10
†Handbook of urban landscape, ed. by Cliff Tandy, pages 59 and 60

watersheds between one zone and another. By careful selection and modification, the sequence of zones which make up the total space of the area can then be defined, and their character, from 'extroverted' to 'introspective', identified. Barriers between zones, links between them, and 'windows' out to the surrounding scenery can be precisely determined. These plans can then be summarised into a single diagram identifying the visual zones into which the site can be divided, indicating their character, and the features which link or separate them, **7**. The method may seem cumbersome, but if the area to be landscaped is large and/or complex, an isovist record, plus notes, sketches and photographs will not only help the designer retain the character of the site in his own mind, but also enable him to communicate it to others during the design process.

Character of surrounding landscape

The character of the landscape or townscape in which the site is embedded also ought to influence the choice of specific techniques from the varied palette available to solve the functional and aesthetic problems in hand. As a first step, it may be useful to determine which (if any) of the following landscape categories suit the particular site.

● *Metropolis* (for instance, inner city areas). Townscape dominated by large-scale man-made elements such as high-rise buildings; intensively-used streets and pavements; elevated motorways and the like. Landscape in such an environment must both look at home visually, and stand up physically to intensive use by uncaring crowds, **8**.

Adding together these two factors (nature of existing townscape and its mode of use) one is forced to the conclusion that the metropolis is *hard-landscape* territory *par excellence*, **3a**. Nan Fairbrother has pointed out that 'no one would furnish a coffee-bar with fragile antiques, or upholster bus-seats in damask and expect them to survive—or even to look attractive. Buses and coffee-bars have developed their new appropriate styles and materials and so must our new public-use landscapes: no untrodden lawns or carefully-edged beds or delicate flowers or shrubs. These belong to gardens, and except in very disciplined communities are too vulnerable. . . . Much ground will need hard surfacing, and in any design the ground-level is especially liable to suffer, whether from over-wear, misuse, litter, or unwanted additions in general. The new style should therefore rely on indestructible effects—on broad massing and spacing, on clear planting of the right trees and robust shrubs, on ground modelling, on the use of light and shade, of enclosure, and open space and vistas. Small-scale effects and intimate details must be provided not by careful maintenance but by hard landscape (patterns of cobbles and setts for instance), and above all by self-renewing vegetation safely above harm's way'*. Whether it is the designer's intention to exploit this theme of bold forms and tough finishes to the utmost, or to provide a well-planned soft contrast (and such a contrast would have to be very well-planned indeed, with all vulnerable greenery amply protected, or situated out of physical reach), this image of man-made topography, large-scale massing, and tough hard finishes provides the aesthetic setting for inner-city landscape.

● *Town* (for instance, towns and inner-ring suburbs). Similarly urban, but the physical scale of buildings, roads and other features of man-made topography will be smaller;

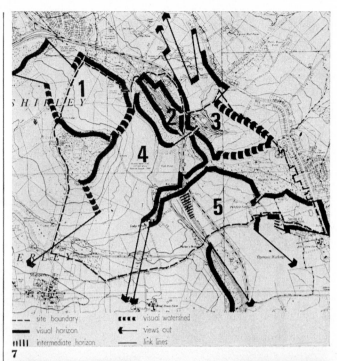
7

8

9

7 *Visual analysis of large complex site by 'isovist' method. Numbers refer to space zones.*
8 *Metropolis—hard landscape territory par excellence (the South Bank complex, London).*
9 *Town—hard landscape predominates, but scale is smaller, character more informal than in 8 (shopping centre in Wolverhampton).*

*New lives, new landscapes, by Nan Fairbrother, page 210

densities of occupation may be lower; and intensities of use less abrasive. While hard landscape will therefore still be very appropriate, and form a dominant component in the townscape, the context will be smaller-scale, less high-powered, and possibly softer than in metropolitan situations, 9.

● *Arcadia and urban parkland* (for instance, outer suburbs and city parks). Low-density occupation, and low-intensity use. Therefore, no longer a dominance of hard landscaping, and probably a preponderance of grass, shrubs, trees and flowers, especially in the case of 'well-disciplined' communities. But there will still need to be extensive reinforcement of the more heavily-used surfaces with stone, brick or concrete, 10, and this can be done without looking incongruous —paved sidewalks and footways, and protected trees and shrubs would be both functionally sensible and visually appropriate.

● *Rural* essentially soft, natural landscape. We have now left the urban domain, and not only is this setting functionally different from the foregoing categories (very low population densities and very low intensities of use), but there is an essential difference in visual character, 11. This is the countryside, and detailing should avoid the vocabularies associated with the city—paved paths and sidewalks, hard upstand kerbs, steel or concrete barriers, and the like—which so effectively suburbanise what ought to be country landscape. Gravel surfaces are preferable to paved footways, and flat kerbs or simple gravel verges are preferable to upstand kerbs.

● *Wilderness* Hard landscape will here be used only in isolated, well-defined elements such as highway surfacing and lay-by detailing, 12, and should not be obtrusive.

Scale of locality

Having assessed the character of the landscape consider next the *scale* of the locality. Man's consciousness and experience of space is an important element in his relationship with his habitat; and it is a useful starting point for the landscaper to distinguish between two fundamentally different orders of urban scale—those spaces *large* enough to be read as spread-out, almost two-dimensional horizontal spaces (for instance the illimitable flatscapes of Brasilia, 14); and those *small* enough to be read as semi-enclosed 'outdoor rooms', 13 (for instance the walled patch of garden, or the tiny urban park

12

13

10 *Urban parkland—soft landscape predominates, but extensive reinforcement of more heavily-trafficked surfaces essential (Grand Union canal, London.*
11 *Rural—sweeping curves of elevated highways fit gracefully into Duisburg countryside, Germany.*
12 *'Mono BG' slabs used to reinforced parking surface.*
13 *Urban scale at one extreme —a town garden at the National Provincial Bank, London.*
14 *The other extreme—a huge floor, stretching away to the horizon (Brasilia).*

boxed in by tall office blocks). Most spaces will of course occupy an intermediate position on this scale, but looking at the extreme cases will help us distinguish between the essential characteristics of the large-scale site and the small-scale site; and from the notes that follow the designer may derive some clues in thinking about the scale of his own particular site.

The scale of any given site will depend on its physical size; on the extent to which it opens out into adjacent open spaces; and on the height, massing and scale of surrounding building façades.

Taking the extreme cases, we can say that very large-scale spaces (such as some of those in Brasilia) are essentially outdoor and two-dimensional in nature: the site has the character of a great open floor, and the designer achieves his effects by means of bold massing arrangements, and the large-scale disposition of 'sculptural' elements, such as trees, clumps of shrubs, hillocks, statuary, and buildings upon this floor. Textures must make sense when viewed from a great distance, as well as when seen from nearby. Small, enclosed patches of space by contrast are semi-indoor and three-dimensional in nature: the site has the character more of an 'outdoor room', **13**, consisting of paved or grassed floor, surrounding façades as walls, and a patch of open sky as ceiling. The designer may see such a site as an extension of the indoor spaces opening out on to it, and will pay a lot of attention to the formation of space by means of enclosing surfaces, and to the small-scale detailing of surface textures.

Proposed landscape function

Having evaluated the nature of the site (and its surroundings) as found, the designer's next step must be to assess the landscaping functions to which that site is to be adapted—and here it is a useful discipline to insist that all external space must have some definite use that has been explicitly defined and carefully thought-out.

Such use may be purely or mainly that of giving *aesthetic* enjoyment, **91**. For instance, a courtyard or flat roof overlooked from surrounding buildings can be planted with greenery and flowers (very easy if plant containers are used), thereby greatly enhancing the quality of the environment for the onlookers. If corridors run alongside courtyard gardens, it is sometimes possible to give occupants almost an illusion of moving through a natural landscape when walking around the building.

But if the site is intended to be occupied and used by people, **15**, as distinct from landscapes meant only to be looked at, there will also be a fairly complex set of *physical* functions to be catered for. There are such general human requirements as comfort, safety and shelter which must be taken into account in practically all urban landscape design: and these are discussed in sections 2 and 3. In addition, special types of urban landscape, such as streets or children's playgrounds, will have special functions of their own (for instance, parking, circulation control, lighting, or the provision of play-shapes); and these are discussed in sections 4 and 5. Finally, the motor-car environment is dealt with in section 6.

1.04 Hard landscape

A common thread running through all landscape design intended for human occupation and use is the problem of wear and tear. All urban spaces (and some rural ones too, such as recreation parks) are vulnerable to the unavoidable wear and tear of intensive use, and that of deliberate vandalism. This problem is getting steadily worse, for population levels are rising and land-using functions such as housing, recreation and travel are increasing, which necessarily means a more intensively used landscape.

Unfortunately, at the very time that intensity of land-use (and,

therefore, the abrasion of the landscape) is increasing, maintenance and repair are becoming prohibitively expensive, and the armies of readily available workers who could be recruited a few decades ago for such fiddly, low-productivity tasks as picking up litter or patching up worn-out lawns, are becoming very scarce indeed.

In this situation there is a clear need for two measures—first, *higher standards* of urban landscape design (which means higher expenditure allocation, and heightened design skills); and second, more *hard landscape* to reinforce vulnerable public spaces against the likely degree of wear.

Concrete is an obvious choice and increasingly used for such hard landscape. Its cost as a material is relatively low compared for instance with stone; and it lends itself to mechanical manufacturing techniques, and efficient site installation—advantages of paramount importance now that craft skills are so rare. But it needs to be used with great care, and a high degree of design skill, if it is to match the attractiveness of traditional materials (brick and stone) both in appearance when new, and in appearance after weathering has taken place. This expertise has in the past often been lacking, with sad consequences for the urban landscape. It is intended that this series will help designers to develop a sensible concrete design vocabulary, and aid them to design urban spaces which will not only be safe, pleasant and comfortable to use, but will also stand up with a minimum of maintenance to the punishing intensity of use associated with urban population concentrations.

15

*Landscaping may be intended for physical use, **15**; or intended primarily to be looked at.*

2 Functions of the pedestrian environment

2.01 Introduction

The pedestrian environment is the connective web of open spaces threaded through the urban fabric, in which people can move about on foot.

The design of these spaces must succeed on two distinct levels—the *aesthetic* and the *physical*—if they are to satisfy users **16**. Section 2 of this series is devoted to an examination of these functions; section 3 will move on to a discussion of the actual physical elements of landscape which may be used to give effect to the functions thus identified.

Aesthetic design is a matter of manipulating spaces, masses, textures and colours in such a way as to produce amenity and delight, and essentially depends on intuitive design ability, sharpened by constant observation and evaluation of real landscapes—constant questioning into the reasons why one landscape is more attractive, more satisfying, and more rewarding than another.

Spatial manipulation in urban landscape is not altogether unlike the manipulation of indoor spaces, a familiar and instructive example of which is the cinema or opera house. A person enters from the open, noisy and probably windy or rainy street into the warmth and glitter of the foyer; he then proceeds through a sequence of dark, constricted corridors and staircases towards his destination, his sense of expectation heightened by the intimate darkness, the disorienting changes of level and direction, and the series of secondary spaces through which he passes on his way (bars, open landings etc). When he finally emerges suddenly into the great auditorium, brightly lit and filled with people, the sense of pleasurable arrival is much enhanced by the dramatic contrast with the previously experienced spaces through which he has passed. There is no reason why the pedestrian moving from a thoroughfare, via a sequence of alleys and staircases, into a great urban square should not be exposed to a similar environmental experience by the skilled manipulation of external spaces. With regard to colours and textures, the main problem in landscape design is to avoid, on the one hand, what Elizabeth Beazley calls the 'bleak, serviceable approach', and on the other hand, the fake picturesque. The golden rule is to produce a varied pattern of textures and colours, derived from the functional differences which distinguish one landscape from

16 *The paving design serves to relate and unify the buildings' changes in level, and planting. In the main, two colours and sizes of precast concrete flags have been used (Golden Lane, London).*

17

19

17 *Small-scale housing at Cumbernauld linked by paved courts and narrow pedestrian ways. The smooth path is of grey standard flags while rough surfaces of bricks and setts protect greenery.*

18 *Sophisticated, formal and precise approach to paving at St Catherine's College, Oxford; in contrast with informality of private garden **21**.*

19 *Well designed signs grouped to form a single 'notice board' and the message is further reinforced by use of coloured paving.*

20 *The 'path' to this courtyard garden in Karlsruhe is clearly defined in precise precast paving while the loose cobbles dissuade pedestrians from reaching the drum fountain and light fittings.*

another, **17**. In other words, *don't* take the line of least resistance and obliterate essential distinctions beneath vast, uniform expanses of grass, concrete or tarmac; but *don't*, on the other hand, introduce fussy, artificial change of material, colour or texture which do not express underlying changes of landscape function. In paving design, for instance, smooth, non-slip, durable hard landscaping, such as concrete paving, can be used on areas expected to bear heavy traffic; a softer, less robust surface, such as gravel, on lightly trafficked areas; and rough, heavily profiled surfaces such as raised cobbles for areas where traffic is to be discouraged, **20**. Surfaces for wheeled and pedestrian traffic, respectively, can be similarly distinguished from each other; and rainwater channels, edge trim, and changes of level and texture can be used to organise

22 *View of the Barbican scheme from Golden Lane estate; the circular concrete drums in the foreground are ventilation shafts for the underground service road and garages.*
23 *Bold and rugged landscape character of domestic garden at Esslingen, Germany.*

22

the landscape geometrically, and introduce patterns and accents, **19**. The colours and textures must always be chosen with both the existing and the desired site character in mind—is the desired character to be formal or informal, **18, 21**; hard and crisp or soft and mossy, bold and rugged or sleek and sophisticated, **22, 23**? The designer must know the answer before he makes decisions.

● The *physical* functions of landscape are the control of wheeled and pedestrian traffic; the accommodation of changes in site level; the provision of climatic shelter; and the provision of protection and security. Each of these functions is discussed in turn below.

2.02 Circulation
Introduction

Aiding and controlling the movement of traffic (both pedestrian and wheeled) is a prime function of hard landscape. Attention should be given to both functional and aesthetic considerations, for a design that frustrates users in either of these respects will be a failure.

Functionally, the problem is one of providing a weather and wear-resistant network of surfaces which will enable people to get from where they are to where they want to be, safely, efficiently, and comfortably; of keeping them out of areas where they ought not to be; of doing all this within the allocated cost. It is particularly important to make the fullest use of low-maintenance surfaces because of the high degree of wear and tear on trafficked surfaces. In practice, this means adequately reinforcing vulnerable areas by means of durable, well-detailed hard landscape, **24**.

Aesthetically, the problem is to make paths which are not only efficient and durable, but which will also look attractive and provide the walker with an unfolding, rich environmental experience. One 'do' and two 'don'ts' may help the designer avoid the sordid pedestrian circulation spaces which are all too common in some cities:

23
→

● Do think of the pedestrian domain as a coherent pattern of places and paths (see **1.02**), with the paths threaded subtly through the landscape, generally skirting around activity or sitting areas, giving pedestrians effortless routes with lots of interest and pleasure along the way, **27**.

● Don't fit paths in as afterthoughts to spaces. Many potentially fine spaces have been cut up into senseless patchworks of landscape, criss-crossed by paths belatedly imposed to follow pedestrian 'desire-lines' instead of being thought out at the outset as part of the total design.

● Don't be intimidated by traffic engineers into pushing the pedestrian into underpasses, or on to windy elevated walkways, just because this suits the motorcar, **25**. The most pleasant and most comfortable place for pedestrians is at ground level, and if anyone has to go over or under the ground, let it be the motorist.

Layout

People tend to take the most direct and effortless path, and if the designer loses sight of this fact, will make their own shortcuts, possibly across vulnerable soft landscape. Therefore layout design should start with a close examination of likely movement patterns between journey *origins* and *destinations* (eg, between site entrance and building entrance; between car park and building entrance; between one building and another; between one site entrance and another).

There are two fundamentally different kinds of pedestrian traffic, and particular designs may have to cater for one, for the other, or sometimes for both.

On the one hand there is fast, purposeful foot traffic (for instance, people going to or from work—moving between underground station and office); on the other there is casual, leisurely, sometimes aimless walking (for instance, strollers in a park, or window-shoppers in a street, **27**). Fast, purposeful

26

24 *Approaches to pedestrian underpass, Düsseldorf. Vulnerable parkland is adequately reinforced by hard landscaping, and pedestrians prevented from crossing dangerous highway by means of clever landscape formation.*
25 *Footbridge across motorway at Zoobrücke is attractively designed, but is bound to be*

exceedingly uncomfortable in windy, cold weather.
26 *London Street, Norwich, where the pedestrian area is enlivened by paving pattern, seating areas, street furniture and planting.*
27 *Casual, dawdling pedestrian traffic needs more space, and interest, than fast, purposeful traffic (Munich).*

27

28

29

28 *Precast concrete exposed aggregate panels have been used to form an upstand barrier to a motorway. Horizontal surface is paved with cobbles hand-set in concrete (Chester Inner Ring Road).*
29 *Sunken barrier (Rome).*

30

31

movement demands direct paths, and paradoxically requires *less* space per person than the slow, casual kind, because people in a hurry tend to move in dense, efficient streams, whereas strollers may walk three abreast, dawdling, stopping and obstructing the way for others. Slow casual traffic also needs more environmental interest than the first kind (for instance shop windows, planting displays, fountains, attractive vistas) and can be led in more roundabout routes in order to expose the pedestrian to changing views and spatial variety.

Clearly then, pathway widths and layouts cannot be decided simply on the basis of origin, destination, volume and density of traffic: the *speed* is equally relevant.

So far we have looked at circulation layout as if on a flat plane, but vertical movement also requires close attention. Most people (particularly if they are elderly or disabled) dislike having to climb long flights of stairs or ramps and often dodge through near-lethal streams of vehicles instead of using the subway provided. Climbing is an effort; subways tend to be sordid; and elevated walkways windy and, in winter, cold.

A fundamental rule, therefore, is to avoid making people go up or down, unless the gradient is easy and short and the environment rewarding. But if there really is no other solution, and pedestrians *must* dive underground, or climb skyward, it is best, by means of positive landscaping, or barriers, to entice them along the desired route, by providing an easy climb and an attractive environment, **24**.

Control elements

Pedestrian flows can be directed thus by three basic methods. These are, in descending order of coercion: by barrier, by hazard and by suggestion.

Barriers

These include upstanding types (walls, screens, fences, and planted beds or hedgerows) and sunken types (ditches, ha-has). Barriers are appropriate where it is vitally important that pedestrians keep to their assigned areas—for instance, alongside dangerous motor roads, **28**.

● Upstand barriers separating motor traffic from foot traffic should be set back the width of the kerb from the road (normally 100 to 150 mm) so that the kerb, and not the barrier, will take accidental impacts from vehicles. Such barriers should have a minimum height of 600 mm, for lower ones could easily cause people to trip over them and fall into the road. Openings in such barriers must be carefully sited to allow drivers to see pedestrians approaching the access points at road crossings. With solid screens, it may be necessary to lower the height of the barrier or to replace it with a transparent structure for a length of 2 m on the traffic side of the opening. Good sight lines are essential for both drivers and pedestrians. Advice on the design and detailing of various

kinds of upstand barrier will be found in sections **3.05, 3.06** and **3.07**.

● Sunken barriers (designed either as hard or soft landscape) can be effective both as physical separators, and as visual features, **29**. In addition, they can serve a drainage function; and while labour cost may be a little higher than for railings, materials cost will probably be lower. Such ditches do not have the visual monotony of long stretches of fencing or railings, and have the advantage of leaving the landscape intact. Their disadvantage is the width of space they occupy (for separating vehicular from foot traffic, a ditch about 1·5 m wide and 1 m deep would be required), and the time and labour required for construction and maintenance. They will be particularly suitable in new developments and improvement schemes; in rural and parkland settings; in municipal parks; in recreation areas; in schools—in fact, wherever the space is available. They can be planted with grass, or seeded with wild flowers to help replace the disappearing habitat of much insect and wildlife.

Hazards

These are a somewhat less coercive form of pedestrian traffic control than barriers, and include knee rails, low banks or walls, rough-textured pavings, and—if soft landscaping is desired—flower beds.

● Knee rails, as already pointed out, should not be used to separate pedestrian paths from dangerous areas, or be used alongside crowded, fast-moving pedestrian ways, because of the danger of people stumbling or being pushed over them. But in informal areas with only light traffic, such as parks, **30**, or for separating suburban pavements from private front gardens, they can be most attractive and have a low visual impact on the landscaping.

● Changes of level may be a 100 mm step, a sloped bank, or a low retaining wall of, say, 600 mm height, **31**. These can be effective methods of keeping people to their allotted areas, but children are often attracted to the idea of walking along the tops of such low walls, which are thus dangerous. And they should certainly not be relied upon to protect pedestrians from fast-moving traffic.

● Deterrent surfaces are those, such as cobbles or setts standing proud of the ground surface or profiled precast pavings, **32**, which are sufficiently uncomfortable to walk on to deter all but the most determined pedestrians.

30 *Knee rails have low visual impact on landscape (Cologne).*
31 *Change of level formed by low retaining wall constructed of concrete blocks (Karlsruhe).*

32

33

34

Suggestion

Control by suggestion is, of course, the least coercive form of control. It includes the use of signals, notices and signs, suggestive changes of texture or level in the ground treatment, and the cunning disposition of landscaping elements generally, to guide people along the chosen route. Preferably several of these methods should be used to reinforce each other, or some pedestrians are likely to miss the message and walk where they ought not to, or lose their way, **19**.

In signs and notices, pictograms and words can reinforce each other. The use of bollards, or mild changes of level such as kerbs, together with suitable changes in paving texture, can serve as suggestive demarcations between areas, without however physically preventing people from crossing the line; and they can add much visual interest to, and help organise, the floorscape, **33**.

Making use of the landscape generally means, for instance, screening from view the directions the walker ought not take, until he has committed himself to the chosen path, and/or providing a focus of interest to draw him in the desired direction, **34**.

Advice on the detailing of various types of pavings which can be used for control by hazard or suggestion is given in section **3.02**.

Conclusion

The decision as to which of the above control elements is most appropriate in any particular situation will depend partly on the function to be served (eg, is it absolutely vital to confine pedestrians to certain routes or areas, or highly desirable, or only mildly desirable?); partly on the visual effect desired; partly on the kinds of people who will be using the spaces concerned (is it a quiet residential area or a tough city area?); and partly on funds. All these considerations should be absorbed by the designer's mind, and what emerges eventually must be an arrangement of spaces, surfaces, textures and colours which will be both functionally sensible and delightful.

2.03 Changes in level

Changes in level can occur either naturally, or artificially, arising out of design decisions; and it is a major function of landscaping to accommodate, and pleasingly sculpt, any changes of level. 'The art of manipulating levels is a large part of the art of townscape' (Gordon Cullen).

Functional reasons for altering existing ground levels include planning needs (ie, the necessity to adjust ground surfaces to the floor and entrance levels of the buildings on site, or to street and driveway levels); circulation control (eg, the use of embankments or retaining walls to guide pedestrians along the appropriate routes and dissuade them from encroaching upon areas where they ought not to be); and visual or acoustic screening.

Aesthetic reasons for altering levels are to introduce visual interest into a dull flatscape by means of mounds and hillocks, sunken gardens, sitting areas, and the like.

Advice on the detailing of hard-landscape changes in level are included in sections **3.04** and **3.05**.

32 *Deterrent surface consisting of precast profiled paving.*
33 *The change in paving colour and texture and the use of bollards provide subtle guidance for pedestrians (Sunderland).*
34 *Visual focus plus linear paving draws pedestrian* *almost irresistably in the desired direction (Wexham Springs, near London).*
35 *Use of concrete umbrellas provide shade (park in Cincinatti, USA).*

2.04 Shelter

Introduction

Shelter from the natural climate is a prime condition for human comfort, particularly in areas where the weather is pleasant for only part of the year. It often requires only a small drop in temperature, a rise in wind, or rain, for conditions to become unpleasant or even intolerable.

Plants too require shelter against sun, low temperatures, or wind (particularly when the latter carries chemical pollution). The provision of adequate climate modification is therefore an important design requirement.

Existing site microclimate

At the outset of design, consideration must be given to the existing site microclimate, that is, the precise conditions of sun, wind, air temperature and so on, within a given area. The designer should visit the site several times in varying weather conditions and, if possible, at different seasons, in order to form a reasonably reliable evaluation. Trying to build up a reliable statistical picture of site microclimate by instrument readings is seldom feasible or even useful.

The following checklist of typical situations which can often be identified, taken from *The AJ Handbook of urban landscape* (by Cliff Tandy) may be helpful:
● Cool north-facing slope with low insolation
● Warm south-facing slope with high insolation
● Bad air drainage (flat low-lying land)
● Sheltered locality (reduction of wind-velocity by physical features such as hills or tree-belts)
● Exposed locality (no reduction of wind-velocity; possibly even a speeding-up as a result, for instance, of funnelling effects of tall slab blocks);
● Tendency to hold fog, mist and humidity

● Particular coastal conditions of off- and on-shore winds, salinity and salt-spray.

Climate modification

With the above information, the designer can decide how to use the form of the proposed landscape to modify the natural microclimate.

Sun

In summer, the problem is to provide shade, 35; and in the cooler seasons, to raise the temperature. In addition to providing direct heat, the sun's heat can also be stored, by providing massive south-facing walls which absorb solar heat during the day and radiate it after the sun has set. These walls will be much more effective where the space is sheltered against the wind, than where it is exposed, for the cooling effect of a chilly wind can easily cancel out the warmth given out by the wall. A wide walled courtyard with low walls on the southern side and high walls on the northern side, or a south-facing corner, to ensure maximum insolation and heat storage, would therefore be sensible. The exact orientation will depend to some degree on site topography and the disposition of buildings—factors often beyond the designer's control. But the time of day when people will be free to use a courtyard ought also to be taken into account. If, for instance, the site is to be used mainly in the morning (this may apply in some schools), a south-east-facing enclosure would be indicated, while a south-west-facing one would suit afternoon and evening users.

To prevent such a sunny space becoming too hot during the summer, trees and wall-covering vegetation may be useful.

Wind

Frequently, wind is the major factor causing external spaces to become uncomfortable, or even intolerable. For instance, on a cool day a fairly small drop in air temperature can often be effectively counteracted by the direct or stored warmth of sunlight in a suitably-designed walled courtyard—but *only* if there is no wind. Wind lowers the effective temperature and in addition causes discomfort by swirling rubbish, newspapers and dust about.

East and north winds are significantly colder than the more frequent south-westerly ones, so it is particularly important that the former be excluded from pedestrian spaces. This can be done by either soft or hard landscaping, as follows.

Trees and shrubs, planted in dense belts, are very effective and their open texture has the advantage of not causing turbulence. In addition, the use of planted landscaping for this purpose may have the advantage of freeing the designer from having to lay out his buildings and hard landscape in possibly undesirable alignments simply in order to provide wind exclusion.

But hard landscape, such as high walls or suitably arranged terraces of buildings, can also be used for wind exclusion, by forming enclosed courtyards, gardens or cloisters, with relatively low walls on the south and west (so as not to obstruct the sun), and high walls on the north and east (for solar heat storage and the exclusion of cold north-easterly winds).

Long solid structures such as high walls or terraces of buildings provide sheltered areas both on the windward side, and even more on the leeward side, **36**. But great care must be taken to prevent turbulence and avoid funnelling effects. This can be done by locating pedestrian areas near the centres of long walls, rather than near the ends, where turbulent conditions tend to occur; by avoiding excessive building height (ie, two- or three-storey terraces are better than fifteen- or twenty-storey slab blocks); and by forming two- or three-sided enclosures rather than straight walls, which, of course, give no protection whatever if the wind is blowing along their length. A cloistered courtyard, in which the height of the enclosing structure is about one tenth the width of the space, is a sensible building form in this respect.

In the final analysis, however, the degree of protection that can be given by such landscaping on a particular site depends very much on the topographical features and air currents of the surrounding area. If there are tall slab blocks in the immediate vicinity, wind conditions are likely to be unpleasant no matter what landscape measures are taken, short of total enclosure. Wind tunnel tests are the only reliable method of investigating situations of this kind. For more detailed information, see AJ *Handbook of building environment* (AJ 27.11.1968 p1283 to 1298; 4.12.1968 p1345 to 1364; and 11.12.1968 p1413 to 1430).

Rain

The problem of rain exclusion is probably well enough understood by most designers to obviate the need for dealing with the subject here.

Noise

Noise nuisance in the open air can be reduced by placing a solid wall solid between noise source and listening position. Such walls should be large, both in height and in length, in relation to noise wavelength and should be close either to noise source or to listening position. A wall or raised bank alongside a motorway can, for instance, substantially reduce the noise nuisance to nearby people, **37**, but such a screen, to be effective, would have to be high enough, and near enough to the motorway (or the listener) to provide a sharp acoustic cut-off angle. Technical details will be found in *AJ Handbook of building environment* (see AJ 22 January 1969 p268); in *Acoustics, noise and buildings* by Parkin and Humphreys p173 (paperback edition); and in *AJ Handbook of building enclosure* (edited by A. J. Elder and Maritz Vandenberg, pp 69-73).

Conclusion

It has been seen that hard landscape features, particularly

wind-protected, masonry-walled spaces exposed to maximum sunlight, modify local climate considerably and in many places may make the difference between comfortable and intolerable outside conditions.

The design of such hard landscape features should not, of course, proceed on the basis of climatic considerations alone. Requirements such as privacy, or the desirability of leaving views clear must be borne in mind; and the requirements of permanence or adaptability also influence choice between massive walls, light screens, or trees and shrubs as enclosures.

2.05 Protection and security

Excluding unwanted people or animals from private property, or from public spaces where their presence might cause damage, is a major function of landscape design. The means used to achieve this protection is usually a physical barrier such as a wall, fence or screen.

But such barriers cannot be designed on functional grounds alone. The type selected has a powerful effect on the visual and social character of the townscape, 38. High masonry walls, pierced screens, picket fences, open knee-rails and bosky hedges all help to create particular kinds of streetscape and townscape. They may symbolise particular kinds of lifestyle. For instance, a green suburban landscape, with only open knee-rails to separate private front gardens from the public sidewalk, will convey to most people an image of a relaxed, neighbourly style of life; whereas the use of high solid walls to protect and enclose front gardens against the street, will probably convey an image of an aloof, private lifestyle. Barriers may also influence social behaviour in subtle ways, for there is reason to believe that in areas of social stress, both vandalism of property, and molestation of people, may be more likely to occur in areas hidden from the view of local residents than in areas under their natural surveillance, and felt by them to be part of their sphere of influence*. Finally, each type of barrier or enclosure may have a particular kind of impact on micro-climate (see section **2.04**).

There can be little doubt that design standards for enclosures of this kind have declined over the past century or so, possibly as a result of the constant preoccupation with cutting costs. Wire mesh is a poor, but all too common, substitute for the fine masonry walls and iron railings of yesteryear. For privacy, shelter, visual quality and low maintenance, a good case could be made for returning to more high-quality masonry walling in housing layouts and other urban landscapes.

On the basis of the degree of security required, and such secondary criteria as the needs of privacy, uninterrupted view, climatic shelter and visual character, and bearing in mind the likely degree of physical abuse (either deliberate vandalism or incidental damage) that may be met with, suitable protective barriers can be chosen from the following range:

● *Ha-ha* (dry or water type): effectively, yet unobtrusively, denies access to the area beyond, while allowing full view across. The water ha-ha can, moreover, make a great visual contribution to landscape.

● *Steep change of level* (embankment or retaining wall): also offers possibility of denying access effectively, but without making the intention obvious.

● *Knee-rail:* allows full view; offers protection mainly by persuasion. Not effective against dogs, or against people who refuse to co-operate.

● *Fence:* retains view while offering more effective protection than knee rail. Design standards are too often inadequate as a result of the tyranny of cost-cutting.

● *Screen wall:* fully effective physical barrier, which also allows some visual privacy and sunlight control.

● *Solid wall:* effective barrier both to unwanted access and to vision; offers sunlight and wind-control if properly designed. Advice on the detailing of fences, railings, walls and screens will be found in sections 3.06 and 3.07 to be published shortly.

* *Defensible space* by Oscar Newan, p79.

38

37

36 *This walled courtyard (Wexham Springs) not only provides protection from the wind but also makes use of the south facing walls for storing heat.*
37 *Existing linear feature (Genoa) provides useful route for the Strada Sopraelevata, minimising both noise and visual intrusion which would otherwise have been caused by the motorway.*

38 *Type of barrier selected for protection and security also has powerful effect on visual character of townscape (concrete fencing).*

3 Main elements in the pedestrian environment

3.01 Introduction

In section 2, we examined in some detail the major *functions* that have to be catered for in the pedestrian environment. In the present section, we move on to a discussion of the *physical elements* of landscaping which can be used to satisfy these functions.

3.02 Paving

Functions

The functions of paving are largely those discussed in section **2.02**. The first is to provide a hard, durable, non-slip surface which is suitable for anticipated traffic and resistant to weather; second, to give pedestrians a sense of direction, guiding them (by layout and pattern) to their destinations, and preventing them (by means of rough, heavily profiled surfaces uncomfortable to walk on) from straying onto areas where they ought not to be; third, to create an attractive and appropriate visual floorscape—one that will express and reinforce the chosen site character (hard or soft; rugged or sleek; formal or informal; etc), and will help link the buildings on site in an appropriate relationship.

Design

Factors to be taken into account in paving design are:

Ease of movement
This is a primary requirement; and once the designer has decided a layout of paths for pedestrian movement (see section 2.02), he can encourage people to follow these paths

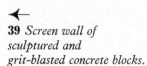

39 *Screen wall of sculptured and grit-blasted concrete blocks.*

40 *Concrete stepping stones provide the obvious route across this water garden in Oregon, USA.*

41

42

by providing comfortable, attractive surfaces, **41**, along these chosen routes, and difficult, uncomfortable surfaces (either hard or soft) to discourage them from taking undesirable shortcuts, **42**.

Safety
Surfaces must be non-slip in both dry and wet conditions. Also, slopes and drainage falls should not be so steep as to pose hazards if encountered unexpectedly (for instance in the dark).

Costs
A prime factor in choosing between one surface-covering and another and an important point to bear in mind is that the true cost of a landscape finish is the initial cost plus maintenance or replacement cost. While grass and gravel are much cheaper than concrete initially the cost of upkeep will probably be so high, in areas of intensive use, that they may ultimately be more expensive (in addition to probably looking threadbare and worn-out).

Appearance
This includes colour, scale and texture. With regard to colour, the main problem designers seem to have when using concrete pavings is to find the golden mean between a boring dullness on the one hand, and a gaudy brightness on the other. Coloured cements should be used with great caution; small differences in colour, shade or texture are usually quite effective in producing variety in surface appearance, and it is undesirable in climates such as that of the UK to resort to bold contrasts and bright colours. Variations in colour or texture should usually be introduced only to reflect functional differences—for instance, to mark suggested pedestrian routes and encourage pedestrians to move in particular directions, **41**. When designing such paving patterns, it is important to bear in mind from which vantage points the pavings will be seen— whether from ground level only, or also from tall surrounding buildings. Patterns should make sense and look attractive to *all* viewers.

Considerations of scale can influence the selection of colour and texture and the design of joints. It is important that the sizes of paving units, their colours and textures be correctly related to the scale of the site. A rough-textured finish on a paving slab may be visually lost in a large area and may be too coarse for a small area. Similarly, the use of coloured aggregates or concrete may be charming in a small-scale scheme but the final colour and effect may be very different when seen in large areas.

The use of joints can be very effective in expressing scale, **43**. In a very large area, a line of paving slabs may be used to provide scale by forming a joint between insitu concrete panels. Or, where smooth paving slabs are the predominant finish, a larger, bolder scale can be provided by flush-jointing a group of four, six or eight slabs together, and surrounding them with a 10 mm recessed joint. In very large areas, bays of paving slabs can be surrounded by 300 mm wide bands of other paving slabs of a different colour or texture as a joint. Split concrete blocks, paviors or setts can similarly be used to

41 *Paving slabs exploited to fulfil the double purpose of providing a comfortable, durable traffic-bearing surface, while also giving pedestrians a sense of direction in a rather ambiguously shaped space, guiding them to their destination.*
42 *Circular paving forming path across a soft-landscaped area.*

43a, b *Joints used to impart a sense of scale to large areas of concrete paving. In **b**, the joints serve also to integrate the paving with the soft character of the surrounding landscape, while in **a**, joint design is hard and crisp, helping to emphasise the contrast between the hard landscaping and the flanking gardens.*

43a

43b

44, 45   46

form joint lines, **44**. In addition, 10 mm recessed joints can be used between single paving slabs to emphasise the scale and to provide a linear or grid effect.

Material and component selection

Paving slabs are available in rectangular and square shapes, **46**, the dimensions of which are given in Information sheet 1. Some manufacturers also supply round or hexagonal interlocking slabs, **47**. Hexagonal slabs give a strong joint line, but are less directional than rectangular slabs, and they are therefore suitable for meandering paths and static areas. Circular slabs, **45**, may be difficult to use because of the awkwardly shaped interstices they create. One solution is to use specially manufactured infill slabs to link them, another to form insitu concrete or gravel surrounds. They give a highly decorative, and rather luxurious, quality to the paved area, and may be suitable, for instance, for restaurant terraces and similar areas.

Finish and texture must be selected to match intended use, and the texture must be able to resist the amount of likely wear. For instance, a soft aggregate may wear through too rapidly, and pea shingle and some aggregates may be too slippery for non-horizontal use. Some aggregates, when exposed, may be too abrasive for foot traffic, whereas an extra-grip surface will be required for gradients.

Future changes due to exposure should be anticipated. This involves more than the effects of just rain and atmospheric pollution. The rate of rainwater run-off controls the cleanliness and collection of dirt. Motor and industrial oil, or petrol drips, can be a nuisance, but where smooth-surfaced pavings are exposed to the rain these can be washed out over a few weeks. With severe exposure, and where there is a likelihood of standing water and concentration of de-icing salts, the use of an air-entrained concrete will increase the resistance to frost. In applications where there is a tendency to flooding (for instance, car washing) the design of the joints and drainage should be given particular consideration. As yet, no-fines paving slabs are not manufactured, but there is undoubtedly considerable scope for their use. No-fines slabs would be porous and could be used, with a suitably designed base, to drain direct to the ground below.

Colour

The use of a single colour can be overpowering and may appear artificial; what is required, if coloured pavings are to

44 *Precast concrete paving modulated by joint thicknesses and setts. (Hamburg, Germany).*
45, 46, 47 *Concrete pavings are available in a variety of shapes—square, hexagonal, circular.*

47

be used, is a carefully considered tonal study—particularly in hazy, rainy climates such as that of the UK where (unlike the Mediterranean areas) bright colours simply do not look acceptable. A final point: if coloured slabs are selected it is wise to first see a sample which has been in use for some months, or which has had the surface laitence removed.

There are three ways of introducing colour to pavings. The first is to use a coloured cement; the second is to use a coloured sand and aggregate; the third is to apply a covering of coloured aggregate to the surface before the concrete sets.

Coloured cements (or white cement) greatly extend the range of possible colour combinations when substituted for ordinary Portland cement. The use of white cement is to be encouraged, as the colour is natural to the material and it can form a very clean and effective setting for light-coloured aggregate, or alternatively a contrast to dark-coloured aggregates. Colours are achieved by colouring either white cement or the ordinary grey Portland cement, or a combination of the two, to obtain the shade required. Some of the colours available in proprietary pavings are the following: slate blue; lilac blue; light green; middle green; grey green; marigold; brick red; and yellow. Black cement can also be obtained. Each precast manufacturer usually restricts his colour-range to only a few colours, and these are normally available only in flat, cement-faced or mildly textured, non-slip slabs. On the other hand, the range of coloured finishes with insitu concrete is all but limitless and very much depends upon the skill and wishes of the designer. When coloured pavings first appeared on the market, the colours tended to be either too strong or too dull, resulting in hard landscapes either dull or gaudy; but colouring and textures have since improved and a moderate and restrained use can produce pleasing results.

Coloured sands and aggregates are the second method of introducing colour into pavings.

Highly attractive effects may be obtained by removing the outer skin of cement and fine material which normally forms on the surface of concrete, thus exposing the aggregate, **48**. Aggregates which are suitable for good concrete are many and varied, and among them may be found a range of attractive colours which look particularly pleasing when exposed.

Numerous methods of exposing the aggregate are in use but the cheapest and simplest method for both insitu and precast concrete is to brush the concrete while it is still sufficiently soft. Stiff bristle or wire brushes should be used, with plenty of water, to remove the matrix and clean each piece of exposed aggregate.

An exposed-aggregate finish revealing the true nature of the material and colour can be provided in the final finish by careful choice of aggregate and cement.

The finished appearance depends on the type, size and shape of aggregate and the colour of constituent materials.

The application of a covering of coloured aggregate is the third method of introducing colour into pavings.

Unusual exposed-aggregate finishes can be achieved with selected aggregates which are scattered on to the finished cast surface and then tamped and trowelled. With smaller-sized aggregates, the surface can be so well-covered that the colour of the cement used in the mass of the concrete is of little consequence. However, in certain instances, the use of a separate facing mix may be appropriate. Provided that the compacted concrete surface is levelled accurately before the chippings are applied, no further treatment is required after trowelling is complete. This technique is usually limited to small chippings, which must be clean and free from dust.

After the aggregate has been fully embedded into the surface by tamping, rolling or trowelling, it is then exposed by spraying with water (see Information sheet 13).

Texture

The texture of both insitu and precast paving can be smooth, profiled or textured. The choice depends on anticipated use, desired appearance (from both near and afar), and the other factors already outlined in earlier sections of this Handbook.

Smooth finishes can be of various types, dependent on the mould material used. Smooth impervious mould materials, such as the plastics and metals, produce a polished finish with the occasional blow-hole, and very smooth finishes such as these tend to craze when subjected to weathering. Moderately absorbent materials (such as moulds lined with specially prepared paper) produce the best finishes, textured enough to avoid undue crazing and without blow-holes. Absorbent mould materials (such as timber, which might be used for specials and one-off designs) have to be treated to reduce the amount of absorption, otherwise the quality of the concrete surface suffers.

Profiled finishes are produced by casting the slabs in moulds with indented faces. The pattern produced can be geometric, linear, circular or dimpled, recessed or standing proud of the surface. Mould patterns may be made from rubber, wood, hardboard, metal or laminated paper, and the material used determines the number of casts which can be made from each mould before it needs replacement. Thermo-setting, vacuum-drawn plastics are suitable for individual designs or for reproducing impressions of other materials such as hessian or stone. When pavings are hydraulically pressed, the range of profiles is limited, but with vibrated pavings there is an ever-increasing variety of moulds available, as the profiled moulding sheet or panel can be placed at the bottom of the mould, and removed afterwards.

A different and very useful kind of profiled slab is the type which has inserts (eg, rubber) standing proud of the surface.

Textured finishes for both insitu and precast concrete paving can be produced in several different ways. For instance, the surface can be scored, **50a**, rolled, **50b**, brushed, or grooved, **50c** and these operations should be carried out while the surface

48 *Exposed-aggregate finish achieved by spraying with water and brushing off the surface layers with a soft brush*

is soft enough to be worked. The extent of the impressions that are made on the surface vary with the degree of set reached by the concrete.

● Scoring can be in random lines along or across the slab or the lines can be set up formally at exact centres, **50a**. The line produced may vary from a simple indentation to the formation of a jagged ridge of cement alongside a rough groove. The depth of surface lines and the size of aggregate both affect the type of line created (see Information sheet 14).

● Rolled finishes are obtained by using a simple roller across the surface. The best results are obtained if the roller is as wide as the slab, for then the whole slab can be covered in one motion. Grooved rollers produce a linear pattern, spiked or dimpled rollers create holes at random, and diamond patterns can be produced by covering the roller with one of the many forms of expanded metal, **50b**. A plain scaffolding tube produces a roughened texture (see Information sheet 14).

● Brushed finishes can provide a non-slip surface with linear or circular ridge effects which depend on the way the surface is brushed. Brushing the surface only raises the surface fines to form the texture, and it is possible for the surface to look patchy due to the aggregate showing through in places. A brushed texture is almost pure sand and cement, and tends to be worn flat by pedestrian traffic (see Information sheet 14).

● Exposed aggregate is produced by removing the surface layer of cement either while the concrete is still green, or after it has set, to expose the aggregate below. It should be re-emphasised that not all aggregates are suitable for paving, and soft, flaky or smooth round aggregates should be avoided—except, possibly, if the pavings are to be used in areas which will not be subjected to heavy traffic, possibly for mainly decorative use. This applies particularly to round smooth aggregates such as pea shingle, which can be slippery and dangerous.

Methods for exposing the aggregate while the concrete is still green include brushing with a wire brush; or brushing with a soft brush and washing down; hosing down; treating with a felt float or scraping (see Information sheet 9).

Methods for exposing the aggregate after the concrete has set include the use of power tools and abrasive blasting. Size and depth of cut can be varied by the choice of tool, and patterns can be produced by tooling or blasting to different depths—in this way lettering, arrows, numbers and signs can be engraved on paving slabs.

Finally, it should be mentioned that existing, already-laid pavings can be textured by tooling or blasting (as above); or by cutting in grooves, **50d**, with a surface planer; or cutting in circles or other patterns.

Drainage

Rainwater run-off is a problem requiring careful thought in designing large areas of paving.

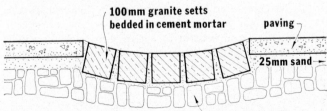

49

49 *Granite setts used as surface water-collecting channel in paving.*
50a, b, c *Three methods of obtaining textured finishes while the concrete is still plastic.*

a *by scoring,* **b** *rolling or* **c** *tamping with steel comb. Alternatively* **d**, *grooves can be cut into already-hardened concrete. This provides excellent anti-skid surface.*

50a

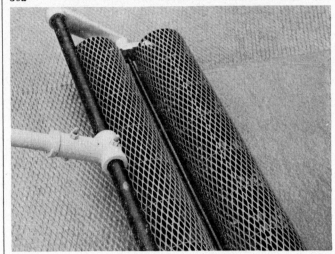

50b

50c

50d

It is theoretically possible to solve the problem of run-off by using porous concrete surfaces, which allow rainwater to percolate down through the slab, and drain away into the ground. No-fines concrete has been used with great success as self-draining surfaces for tennis courts, and the wider application of this material is now considered. No-fines could be used for the manufacture of paving slabs, which would be quite strong enough to withstand the stresses of rough handling, delivery and laying. The thickness required to withstand the loading of pedestrian traffic has not yet been determined, but 70 mm thick no-fines concrete laid on well-compacted fill appears to be strong enough for car parks.

However, the use of such porous concrete surfaces also raises problems. The most obvious is the fact that the sub-base would need to be porous, in order to drain away the water percolating down through the concrete. When concrete is laid insitu this need be no problem, but a porous sub-base would not necessarily suit precast paving slabs.

The second main disadvantage is due to the surface texture of the concrete, which would have larger voids than normal pedestrian paving. It is quite possible that these voids may become traps for dirt and rubbish, and may break up under stiletto heels. Also, if the surface holes are large enough the smaller particles of dirt may wash right through into the interstices below. And finally the surface, while providing an extremely good grip in the dry, may become polished with use and slippery when wet. A possible solution to this might be to surface the pavings with a suitable non-slip aggregate.

If these problems are solved, no-fines concrete would be a good self-draining covering for large areas, with a substantial application for instance in intensively-used all-weather recreation areas. If on the other hand ordinary impermeable concrete pavings are used, provision must be made for directing the flows of rain-water away from the areas used by pedestrians, into channels and drains, so that surfaces are never covered by water. Minimum falls for roads are 1:50; for cycle tracks 1:40; for footways 1:30 to 1:40; and for general paving 1:60 in the case of insitu concrete, and 1:70 in the case of precast slabs. These falls should lead the surface water to collecting points or channels, to be removed ultimately to a sewer, soakaway trench, or natural watercourse.

The integration of drainage installation and pavings can be achieved by the use of matching precast concrete drainage units (or by using granite setts, **49**, or some similar method to form the channel) which can be obtained to match most forms of square, rectangular or hexagonal pavings, with appropriate finishes and textures. Manufacturers can provide rectangular and hexagonal manhole and inspection covers which are flat-topped, with recessed extending lifting handles, to match their pavings. Perforated standard slabs are also available as covers to gulleys, and most manufacturers make specials if required. Dished precast concrete channels are readily available and their appearance could possibly be matched to suit the surrounding paving.

3.03 Junctions and trim

The functions of kerbs and trim are to protect the edges of pavings, and to prevent the base from spreading; to mark the boundaries between pavings and other surfaces; to define areas of paving and mark traffic routes; to mark changes of level; and to form construction joints, **51a, b, c, d**.

Details should be selected not only with an eye to cost and the ability to stand up to the anticipated degree of abrasion, impact and other wear and tear, but also bearing in mind the visual effect that trim can have on landscaping character, **52**. Elizabeth Beazley's point about the suburbanising effect that formal upstand kerbs may have on country paths has already been mentioned (section 1.03).

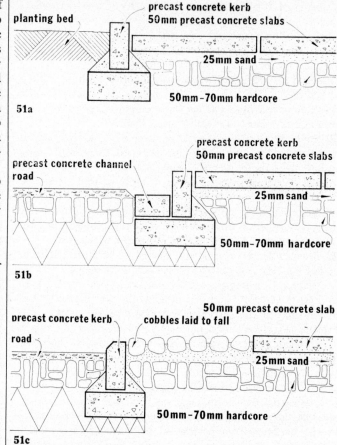

51a, b, c Three forms of junctions and trim: **a** junction of precast concrete paving and planting, using standard precast kerb, **b** junction of precast concrete paving, and road, using standard precast concrete kerb and channel, **c** junction of precast concrete paving and road, using standard precast concrete kerb and separating border of cobbles laid to fall to road.

Kerbs and edge trim

Change of level in pavings is one way of informing the pedestrian of the proximity of a vehicle route. A narrow strip 300 mm wide, 70 mm above or below the paving level should provide adequate warning. But both these measures could be a source of danger, causing people to trip or stumble. The alternative is to provide a slope, either downward or upward. A downward slope from the pedestrian surface into the road could cause rollerskates or pram-wheels to swing into the road; therefore an upward slope appears to be the best answer. A slope leading up to the kerb about 300 mm wide at an angle of about 10° would be quite safe, would provide a restraint to pram-wheels, and would provide a physical warning of the proximity of the road. Cut-outs would have to be formed in the upstand kerb to allow surface water to drain off to the road at frequent intervals. As an alternative to a change in level, a mildly ribbed surface could be incorporated along the paving edge to provide a change of texture; but drainage might then be a problem.

Junctions between pavings and walls

Where paths meet walls or similar solid vertical faces at right angles, the junction so formed is a potential dirt trap. Any dirt or litter which settles on the path is scattered towards the edge by the passage of pedestrians and by the wind. The result is that all the dirt and litter collected on the path surface tends to end up against the wall (or in the gutter). If a 45° splayed base were provided to the wall, the junction would be easier to clean, and it would also be subject to washing by rain, **53**.

51d *Standard precast concrete kerb used to form surround to rock and water garden.*
52 *Choice of trim should harmonise with landscape character. In this German example, 'BG' slabs have been used for demarcation between road and cycle track.*
53 *Carefully detailed concrete slope ensures good appearance and weathering (Tupton Hall comprehensive school).*

51d

52, 53

3.04 Steps and ramps

Primarily steps and ramps have a physical function—that of enabling pedestrians to get from one level to another. But they also have a tremendous potential for heightening the environmental character of a site; and this potential should be exploited where possible, **55** to **57**. Steps can be narrow and intimate, **56**, or wide and monumental, **57**; enclosed and secret, or open and sweeping; and they are an excellent device for luring a walker to his destination with a dramatic sense of gradual approach and climactic arrival. The goal may either be brought into view gradually as the climber ascends a flight of stairs; or it may be tantalizingly displayed as he descends towards it; or it could be revealed with stunning suddenness as he emerges from an enclosed 'secret' staircase.

Deciding the most appropriate material and physical detailing for the steps will depend partly on the desired site character. Elizabeth Beazley* identifies three characteristic types of steps; the rugged, elemental, almost primitive type, giving the appearance of having been carved bodily out of the site; those which fly daringly over the site, heightening the climber's sense of defying gravity as he divorces himself from the earthbound; and those forming part of the architecture on the site, rather than being part of the landscape.

In the first, carved-out type, unity of material is important. For a really rugged, elemental effect, both risers and treads ought to be faced in the same material as the site itself (eg insitu concrete; or concrete paving). On a site with a natural rocky surface, the aggregate used in the concrete may be taken from the rock on site to promote this sense of unity. For a rather more sophisticated, urban character, the designer may face the treads with the same material as the site paving (say concrete slabs), but use a different facing for the risers.

*Design and detail of the space between buildings, page 218

The second kind of stair, which emphasises its defiance of gravity rather than its earthbound unity with the site, can be formed by cantilevering precast concrete treads from a side wall.

The third kind, belonging to a realm of architecture rather than landscaping, will not be discussed here.

Dimensions and gradients

The following information is taken from the *AJ Handbook of Urban Landscape* (edited by Cliff Tandy):

Steps

Risers should be between 80 mm and 160 mm, and goings not less than 300 mm. Tread projection should never be more than 15 mm, to avoid danger of tripping. Satisfactory gradients are between 1:2 and 1:7. About eleven steps make a comfortable flight length, with about 19 steps the maximum. Width of lands should not be less than a metre. Various constructions and details are shown in **54a, b**.

54a *In-situ concrete steps. A non-slip finish can be provided on the concrete before it hardens by brushing with a stiff broom before hardening, by hand tamping, by the addition of carborundum to the top surface or by inserting strips of non-slip material.*
54b *Precast concrete treads and brick or granite sett risers on a concrete base.*

Treads should be 50 mm thick.
55-57 *In addition to physical function of getting pedestrians from one level to another, steps can also make enormous contribution to environmental character, as exemplified by these contrasting examples.*

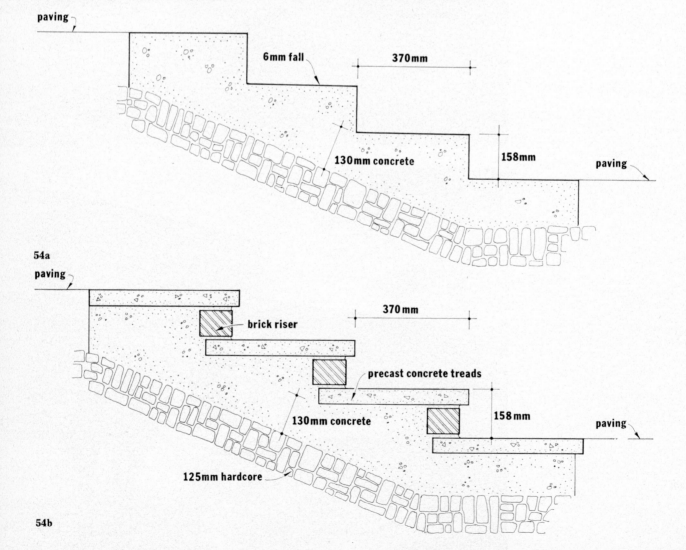

paving

6mm fall
370mm
130mm concrete
158mm
paving

54a

paving

brick riser
370mm
precast concrete treads
130mm concrete
158mm
paving
125mm hardcore

54b

55

56, 57

58

59

58 *Effective use of both ramps and steps at Sunderland civic centre.* **59** *Concrete pavings lapped in random fashion to provide access to church near Copenhagen.*

Ramps

For short distances a pedestrian ramp, **58**, may be as steep as 1:6·5; for wheelchairs and prams, ramps should not be steeper than 1:10, and 1:12 is preferable. Surfaces must be non-slip, and surface water should be shed across the width of the ramp, instead of down the gradient.

Ramped steps

On long ascents, ramped steps should be considered, with three or four steps between ramp sections. If for perambulators as well as pedestrians, gradient should not exceed 1:12, and risers should not be higher than about 100 mm. Nosings must be clearly defined to ensure that users see the steps.

Details of construction

Precast concrete slabs can be used for steps of all kinds—straight, curving and cantilevered.

They need not be used only horizontally, and have great scope for exploitation in three dimensions. There are five basic considerations from which limitless combinations can be developed:

● One slab can be lapped over in any direction, not simply parallel to it **59**;

● An 80 mm slab supported securely at the corners and centre by bricks or other slabs can carry the load of all pedestrian traffic;

● A slab can be embedded vertically in a footing with a free projection of up to $\frac{1}{2}$ or $\frac{3}{4}$ of its length. If it is not liable to be damaged by vehicles or bicycles it can be used in any location provided it has lateral support such as another vertical slab, or where another slab overhangs it horizontally.

● As slabs may have rough edges, it is important to avoid potentially dangerous situations.

● Frequently both faces of paving slabs are quite presentable, and can be exploited where necessary.

3.05 Banks, slopes and retaining walls

Banks, slopes and retaining walls are methods of changing level on site. The need for changes of level may arise either from natural contours of the site, or from a deliberate decision to amend these contours for functional reasons (planning needs; circulation control; visual or acoustic screening) or aesthetic ones.

Whatever the reason for their introduction, such variations in level provide a major opportunity for the sculpting and moulding of the landscape. Many of the most pleasing urban environments in existence (hillside towns, or harbour villages) derive their powerful sense of place, of spatial richness, from the artful manipulation of levels.

Banks and slopes

Because they are tilted up to face passers-by these often make a more significant impact on the townscape than horizontal pavings. They therefore require careful handling of surface textures, and careful detailing, particularly along the top and bottom where ragged edges may all too easily result from careless design.

Concrete pavings are very suitable for facing slopes, especially the hexagonal types with their attractive interlocking patterns; and there is considerable scope for the use of exposed aggregate and textured surfaces. Laying techniques will vary with the angle of the slope. Where slopes are shallow enough for the weight of the slabs to be carried by the ground and by friction, normal laying procedures as for horizontal paving should be satisfactory. With steeper slopes the weight of each slab will partly be carried by the slab below, and greater care should be taken with the compaction and laying of the backing fill. In all cases there should be a concrete base at the foot of the slope, the size of which will depend on the angle of slope, the area of paving, the soil type, and any adjacent structure to which the load can be transmitted. A structural engineer should be consulted when in doubt.

An important consideration is the effect of thermal movement on large sloping areas. When the slabs expand they will tend to creep uphill, and on cooling will return downhill, so that cracking at the joints may result. The joints may also crack due to soil movement, as soil dries out in hot weather. Any small cracks will rapidly deteriorate when exposed, due to the action of rain and weather. The rapidity and volume of rain-water-flow down the face of the slope will tend to erode the joints, and if water manages to penetrate through a joint it may undermine the whole bedding of the slabs. Joints should therefore be watertight and durable.

Retaining walls

When slabs are used as low retaining walls the angle of the retaining wall should allow a safety margin for the weight of the slab to override that of the material restrained. The slabs should rest on a concrete base similar to that for kerbs, with at least the bottom 25 mm bedded in concrete. The bottom slab should be haunched behind, with concrete carried up to at least half the height of the slab. The joint should be pointed, preferably from behind to avoid weeping. The slab will restrain the material retained partly by its design, mostly by its weight, and partly due to the suction of the haunching and of the backing material. The top edge should not be exposed to outward knocks or for people to pull at. Slabs can in theory be tied back with wire wall ties, but this may be difficult in practice. Always ensure that the backs of slabs are clean and offer a good key to the haunching.

3.06 Fences and railings

The functions of fences and railings are those discussed in sections 2.02 and 2.05—guiding pedestrian traffic along the appropriate routes; preventing intruders from entering private property; and protecting pedestrians from hazards such as water, steep falls or vehicles.

In addition, as has been pointed out, fences make an important contribution to the visual quality of the landscape, **60**. The decision to have a fence, and the type of fence chosen should

60 *Precast concrete fencing at Pitlochry, Perthshire.*

be carefully considered because fences are often just as conspicuous as the buildings they enclose. Traditional fences employing local materials and craft practices often achieved a harmony with the landscape, whereas today cheapness of short-life materials and their wide availability has all but eliminated such regional characteristics. Some authorities have attempted to preserve the landscape character of local areas when affected by road improvements, by insisting that wherever possible the new boundaries should be built of local materials. The most acceptable fences, regardless of material, are often those of simple design devoid of contrived effect. In open country, fences should follow ground contours, avoid fussy steps and skylining so that they are subordinate to ground form and tree groups, **62**. The resolution of ground levels at the feet of fences is seldom given sufficient consideration and this is usually due to the use of small-scale plans, devoid of information on levels and detailed instructions for the erectors. Adjoining fences performing similar functions should not be in different materials or colours; and gates should be designed to match the adjoining fence in character.

Fence types

The widespread use of ill-designed concrete-post and chain-link fencing, as a cut-price method of enclosing property, has unfortunately tended to give concrete a bad public image as a material for fencing; but attractive forms of concrete fencing are available.

A bold and robust fencing type, with a strong horizontal visual line, is the concrete post and panel fence, **61**, **62**. This consists of concrete posts grooved to receive concrete slabs slotted in one on top of the other, to form infill panels. Various heights are available.

Concrete palisade fencing is similar in form to traditional timber palisade fences, except that posts, rails and palisades are all precast in concrete, with concealed fixings.

In both cases the appearance of the fence depends very much on the detailed design, and quality of manufacture; therefore it is recommended that a selection of types be inspected in person (instead of relying on verbal description or drawings), and the preferred type or specified by name.

3.07 Walls and screens

The functions of walls and screens in the landscape are those discussed in sections 2.04 and 2.05—to act as physical barriers against intruders; to act as visual barriers for privacy; to act as wind and noise barriers; to help raise the effective temperature in outdoor spaces by absorbing, storing and re-radiating the warmth of sunlight; and finally to define visual space and help create particular townscape character.

Concrete walls and screens have advantages in terms of cost, durability, and their ability to harmonise visually with the rest of the landscape if concrete is also being extensively used for pavings, plant containers and the like. But they require careful material and component selection, and constructional detailing, if they are to be visually appealing, **63**. Both grey dullness, and gaudy brightness should be avoided (see the notes on *Colour* in section 3.02); and very careful attention should be given to weathering and rainwater throw-off, to avoid the sordid streaking and staining effects which are so disliked by the general public.

61 *Example of successful concrete fence, blending with landscape formation and avoiding abrupt steps and fussy detailing.*

Precast construction

Considerable developments have taken place in the use of concrete blocks for external walls in recent years. At present, it is not yet generally understood that recent developments in blockwork are such that they no longer require external rendering. Although there has been considerable experience on the performance of concrete blocks for external walls this knowledge is not yet widespread.

Sizes of blocks and the different bonding patterns available are so numerous that great care is required to limit them and so avoid chaos; this requires the closest possible co-operation between manufacturers, architects, and builders. A similar close partnership between architects and manufacturers is required to improve the appearance of some existing blocks and to develop new forms of surface.

Machine made mass-produced blocks of high quality offer an interesting range of colours and textures including the possibility of vertical striation or grooving. Further scope is available to the designer in the choice of bonding pattern and the method of jointing. Special quoin blocks and reveal blocks are usually available as standard items, but careful planning at the design stage, to avoid the use of large numbers of specials, will be amply repaid by simplicity and speed during construction.

Other types of blocks with special sculptured, 63, or exposed aggregate faces are available from some manufacturers. Design of walls using these blocks involves the same considerations as for mass produced blocks.

Fair-faced blockwork may be painted and such paints should be alkali resistant; if not; an alkali resistant primer should be used first.

Sample panels

Regardless of which finish is selected, it is recommended that sample panel walls be constructed on all important jobs as a means of conveying to block producers, block-layers and building contractors, grade of materials, quality of workmanship and precise appearance desired in the finished walls. Such panels are much more effective than written words or photographs in defining and specifying the quality of work required. Panels should be at least 8 blocks long and at least six courses high and constructed of block of the same type, size and surface texture as will be used in the finished wall. The units also should be laid in the same pattern with the same type of joints required in the finished work. If finished walls are to be painted so should the panel. Panels should be approved by the architect and left intact for reference until construction is completed in a satisfactory manner.

Bonding patterns

Many wall patterns are possible. The units may be laid in regular courses of the same height, or in courses of two or more different heights, or several sizes of units may be laid up in a prearranged ashlar pattern. In some wall treatments all joints are accentuated by deep tooling; in others only horizontal joints are accentuated. In the latter treatment, vertical joints after tooling are refilled with mortar and then rubbed flush after the mortar has partially hardened to give it a texture similar to that of the concrete masonry units. In this treatment the tooled horizontal joints stand out in strong relief. Where an especially massive effect is sought, every second or third horizontal joint may be tooled, with all other joints, both horizontal and vertical, rubbed flush after tooling and refilling

62 *Good appearance depends not merely on specifying correct fence type, but also on getting the detailed design right, and on good workmanship.*

63

with mortar. Another variation in finish is obtained by
extruded joints where excess mortar is used. Some of the
mortar is squeezed out or extruded as the blocks are set and
pressed into place.

This mortar is not trimmed off but left to harden in its
extruded form.

Interesting bonding patterns can be obtained by projecting
some of the units in block wall.

This can be done in two ways:

● By use of blocks of the same thickness as the wall, laid so
that they project beyond the face with a corresponding recess
on the other face.

● By use of deeper blocks which project from one or both
faces of the wall. This is preferable if stability is not to be
reduced. When units are projected they should be spaced so
as not to encourage children to climb the projections.

Stack bond is the quickest and therefore most economical
bond to lay. It eliminates the need for half blocks at the ends
of the wall and so reduces the labour required in cutting units,
handling and storing half blocks. It may be advisable to
reinforce alternate horizontal bed joints in large panels or
when the wall is subject to loading.

Interesting patterns can also be achieved by varying course
heights, by the use of blocks of different vertical dimensions.
Basket-weave patterns are possible provided there is a suitable
relationship between the length and height of blocks. There
may, however, be some danger in using cored blocks on end,
and solid blocks are to be preferred. Particular care is necessary
in the bonding at intersections of walls so as to reduce cracking
due to restraint.

It is important that bonding of blocks should be such as to
distribute loads uniformly throughout the wall. Normally
the horizontal distance between perpends in adjacent courses
of blockwork should not be less than a quarter of the length
of the units.

In choosing the bonding pattern it is important to consider
the dimensions of the panel and the effect of any openings.
Because of the wide range of available shapes and sizes of
blocks, the variety of bonding patterns for facing blockwork is
unlimited. However, those patterns which depart from the
requirements for stability should not be used unless experience
or experimental data indicates that they are satisfactory for
the particular construction. Steel reinforcement in bed joints
will tend to distribute loadings throughout the wall and where
this is used the rules of bonding may be relaxed to some
extent.

While any of the various types of joints may be used with
concrete masonry, either concave or v-shaped joints are
preferred. They provide a tight joint with good appearance.
After joints have been tooled, any mortar burrs should be
trimmed off flush with face of wall with a trowel.

In situ construction

In addition to the precast products previously mentioned, the
plastic nature of concrete either in the form of composite or
in situ construction provides limitless scope and potential for
design.

63 *Concrete network formed
by using concrete screen
walling units.*

4 Street furniture

4.01 Introduction

Street furniture has a significant visual impact on the urban landscape, **64,** and design standards in this regard have long been inadequate—both the design of the individual items themselves, and their grouping and incorporation in the streetscape. Because a multiplicity of statutory, commercial and private undertakings is responsible for the installation of these different objects, they tend to be designed and sited quite independently of each other and the resulting clutter not only looks untidy but actually interferes with the communication of essential information. The situation would be considerably improved if the selection and placing of all elements of street furniture could be co-ordinated by the landscape designer instead of being left to a series of *ad hoc* decisions.

The aim in streetscape design should be to avoid clutter by the sensible grouping together of items wherever possible, **19**; to avoid using items that are out of scale with each other or the rest of the environment; to use colours that are as neutral as possible, unless there is really a good reason for doing otherwise; and finally to pay very close attention to the detailing of the junctions where units are fixed to pavement or wall.

Paving is not usually designed to go round such vertical services as lights, bus stops, telephone boxes and letter boxes,

64 *Precast concrete containers used to form buffer zone between pedestrian precinct and shopping arcade in Munich.*

but where it meets such objects, the paving is usually cut to suit. When carried out by craftsmen, such trimming can be almost imperceptible, but the junctions of vertical with horizontal can still appear very abrupt. The solution is to design for some form of base round a vertical object using another, perhaps darker, textured material; or perhaps to use gravel, **65**, and planting where appropriate. With difficult shapes, the best solution is to lay in-situ cobbles all around, **67**. These would then take up the shape of the unit and of the paving.

Every object used involves a colour decision. This has always been true to some extent, but when local materials (stone and brick, slate and tile) were in common use, the problem largely looked after itself. Today, this automatic discipline has largely disappeared, and sensible, sensitive decisions are required. No colour can be judged to be good or bad in isolation from its surroundings. What counts is how it relates to other colours or contrasts with them. Colour can only reinforce or play down an existing situation—it cannot, for instance, make a bad design good, but it can certainly detract from a good design, and it may possibly make a mediocre design a little more acceptable.

Colour cannot be considered in isolation from the texture of the material, which affects its light absorption and reflection. This is particularly true of coloured concrete finishes which are textured and broken up by shadows, for this deepens the tone. Conversely, a smooth surface, particularly if gloss-painted, reflects light and the surface appears to be lighter than it is. The location of a material also affects its light reflection and therefore its apparent colour. Sympathetic colours are usually in the earth range—yellow ochres, terra-cotta red, and browns, including greenish-brown and warm greys. These colours can either be used to blend or to stand out,

by choice of both hue and tone value. Yellow ochres are very useful for drawing attention to objects without seeming garish. Black and white are excellent colours wherever contrast is needed, and more sympathetic than most greens. Green rarely matches a natural background since the natural 'green' not only contains infinite variety but is constantly changing both with the season and with the light.

65 *Round plant containers set in shallow pebble beds—one solution to problem of awkward junctions between vertical elements and horizontal surfaces.*
66 *Timber seats and concrete plant containers designed as coherent grouping of elements.*
67 *An alternative to* **65**: *circular containers integrated into rectangular paving pattern by use of cobbles around the base.*

66

65

67

4.02 Street lighting

Lighting is usually one of the more dominant elements of furniture affecting the character of a street, **68-70**, and therefore requires close attention.

The functions of street lighting are to aid the safe movement of vehicles and pedestrians; to help maintain public order and safety; and to illuminate the urban scene as an amenity. The choice of height, spacing and type of lighting columns will be influenced by the need to serve these functions.

Lighting of city streets is the responsibility of the highway authority (in the UK usually the city council or the county authority). Installation, maintenance and, often, design are carried out under the authority of the city or the county engineer. In the UK, lighting of trunk roads is the responsibility of the Department of the Environment, the local or county authority acting as its agent. In many of the above instances, the city architect or parks department may also be involved.

Guidance on the siting of lighting equipment is given in BS CP 1004: 1963; advice in particular cases may be obtained from the Royal Fine Art Commission (aspects of civic amenity and situations of historic importance); the Council of Industrial Design (design of individual items); and the Association of Public Lighting Engineers (technical questions of public lighting).

Design

Lighting equipment should be selected and placed so that it looks good both by day and by night. It is important to avoid clutter (eg, by using fewer and more powerful units mounted at higher levels); to use columns that are in scale with adjacent building heights; and to use lanterns that are in scale with the columns upon which they are mounted, **68**. The columns themselves should, of course, be neat and trim, **69**. The general aim should be inconspicuousness and harmony, and in some instances it may be desirable to dispense with columns altogether, and mount the lanterns on building façades.

For more information, it is recommended that readers refer to Information sheet 40 of the *AJ Handbook of Urban landscape*, to the COID *Street furniture from the design index*, and to BS 1308: 1957 *Concrete street lighting*.

68 *Example of lantern, and concrete column, which are in scale both with each other, and with surrounding streetscape. Column is five metres tall, and has slate-grey finish.*
69 *Taller column with larger lantern—but design is neat and trim.*
70 *Of all items of street furniture, lighting standards may be most dominant, as exemplified by these ten-metre-high precast concrete units.*

69

68

70

4.03 Seats

Outdoor seats may be provided in situations where location is
pre-determined (such as at bus stops) or where the landscaper
has more freedom to choose siting and arrangement (such as
in casual sitting areas, **71, 72**).

In the former case, seats should not be unpleasantly close to
vehicles or pedestrians and there should be a clear view of
approaching buses. In the latter case, more attention can be
given to the separation of sitting areas from traffic routes (both
vehicular and pedestrian) and to the creation of quiet enclaves.
The latter can be done either by *layout* (ie, by the formation of
quiet alcoves leading off the busy pedestrian routes, **75**) or by
level (the formation of a sitting area either at a higher or a
lower level than the surrounding landscape). A higher-level
sitting area can be chosen for its views, if there is some nearby
attraction such as a river, but it may be uncomfortably
exposed to the weather and require climatic protection.
A lower-level sitting area may have seats grouped round
flowerbeds, pools, fountains and the like to provide some
focus of attention for the sitters. Such sitting areas should be
exposed to the south for insolation, and should be protected
from north and east winds by walls or trees and shrubs. (These
directions apply, of course, only to the northern hemisphere.)
Seats should not be dotted about a site haphazardly. Coherent
grouping of seats, **66**, lighting standards, plant containers and
litter bins creates a better streetscape than scattering these
elements about singly. The robustness of the seat itself
should be commensurate with the visual setting (concrete
seats look better in cobble, stone or concrete landscapes than
on grass surfaces), and with the kind of physical treatment it is
likely to get from users, **74, 75**. For examples of seats, see the
COID publication *Street furniture from the design index.*

72a

71, 72a, b *Informal seating
areas exploiting a variety of
forms—solid concrete with,
and without, timber surfaces.*
73 *Nice integration of seat
with flower-plant box. Seat is*

two metres in diameter.
74 *Timber seat cantilevered
from solid concrete central
support—a relatively
well-designed element in
mostly non-descript space.*

75 *Robust seats which will
take rough treatment—timber
slats supported on concrete
frames fixed to paving.*

71

72b

73

75

74

4.04 Litter bins

Litter bins should be sited at the precise points where rubbish is likely to be dropped (bus stops; sitting areas; vending machines; confectionery and cigarette kiosks; and ice-cream van parking positions) if they are to be effective. To prevent these rather unsightly fittings being too obtrusive they should be grouped with other items (seats; bollards; and lighting columns) and they should be fixed to some larger object, such as a wall, a column or a fence, rather than be left on their own. Not only should streets be provided with many more litter bins than at present, but more generous capacities should be specified if the prevalent problems of overstuffed receptacles and litter-strewn surrounding areas are to be avoided.

The design of the receptacle will be influenced by the anticipated frequency of emptying. If the bin is likely to be emptied daily, it can be lidless, 77; but bins likely to be emptied only infrequently, or used for the deposit of refuse which might quickly decay and attract flies, should have hinged weatherproof covers to avoid causing a nuisance.

Bins must be thoroughly robust, not only in areas where misuse and vandalism are likely to occur, but also in locations where they may be accidentally knocked by vehicles, eg, in car parks or alongside driveways.

Proprietary types are available in precast concrete, metal, timber or glass fibre, or the unit may consist simply of a paper sack held by a metal ring. Selection must be on the basis of cost, robustness and appearance. Concrete types may not look appropriate in rustic or soft-landscape settings, but in hard landscapes (particularly those already incorporating a good deal of concrete paving, etc) they are very appropriate, and stand up to the rough treatment usually associated with those urban or recreational areas where hard landscaping has been thought to be necessary. A removable liner within the outer container makes for easy emptying.

The ground under and immediately surrounding the bin may be expected to become soiled, so it should be given a compacted or hard surface such as concrete and be marginally higher than the surrounding ground for ease of cleaning.

76 *Robust in-situ concrete litter bin, able to take rough treatment.*
77 *Smaller precast litter bin with removable mesh liner.*
78 *Robust concrete bollards with recesses and differences in finish exploited to enhance sculptural qualities of form.*

76

77

4.05 Signs

Much of the visual chaos of the contemporary streetscape is due to the multiplicity of badly-related signs and signals. Not only is the clutter unsightly, but visual confusion makes the various messages less effective.

Such devices cannot be eliminated, but their impact on the urban landscape can be rendered more effective and less confusing, by grouping them together coherently as far as this is compatible with their purpose, by combining them where possible, and by using simple, well-designed signs of uniform appearance and typeface, **19**. In this way, traffic signs can add to the visual liveliness of the street, and foster a sense of orientation instead of creating confusion and disorder. A useful British guide that may be consulted is the report of the committee on traffic signs for all-purpose roads, published by HMSO under the title *MOT traffic signs*.

The aim in locating signs should be maximum visibility and clarity of the whole array. No sign should distract attention from, or obscure, another equally important sign. Mounting heights should be correlated and not vary unnecessarily: signs for pedestrians should be above standing eye height.

4.06 Bollards

Bollards can be used to prevent vehicles from encroaching on to certain areas without, however, obstructing pedestrians; and for visual reasons, eg, to mark boundaries, define areas or provide repetitive vertical accents on the city floor. Usually, of course, the aim is twofold—to guide traffic flows while simultaneously creating an effective townscape, **79**.

Bollards must be robust both physically and visually, and concrete is therefore a very appropriate material, **78**. They must also be strongly fixed. For vehicular control, the bollards should be embedded in a 100 mm concrete base of 1:2:6 mix, usually to a depth of 300 to 450 mm below ground to resist impact. For pedestrian use, the bollards must be stable and secure but they need not necessarily be concreted in. It is possible to obtain pedestrian bollards which are simply the above-ground length, with a hole in the bottom which is located on a metal rod cast into a suitable base. These bollards can be removed for vehicular access simply by lifting them off the 50 mm rod, and they can be easily replaced if damaged. A wide range of designs, sizes and finishes are available, and these should be chosen to suit the scale of use and the adjacent elements. Bollards with a reflecting aggregate which is clearly visible in car headlights at night are produced by several manufacturers. Although only a few firms manufacture standard bollards, many others will make them to special designs. It is difficult to give design rules, but if a bollard height of one metre above ground is required, a minimum diameter at ground level of about 300 mm will probably be necessary to give the required visual robustness.

78

79 *Simple cylindrical concrete bollards with exposed finish form an attractive line of demarcation between pedestrian and vehicle zones.*
80 *In addition to their function of traffic control, bollards can also form effective vertical accents on paving pattern.*

79

80

4.07 Plant containers

Plant containers have a number of uses: they can be used in temporary arrangements, eg, for festivals or exhibitions; in situations where re-arrangements may be necessary from time to time; in low-maintenance or no-maintenance situations, where annual replacement of plants is preferred to the alternative of constant care; and they can be used to define spaces and paths in a way that imparts a particular desired character to the landscape.

They usually look most appropriate in situations where plants cannot be grown naturally, that is, on gravel or concrete paving, rather than on lawns, **64-67**. They should not be placed near fumes, or in inadequate light and should not be dotted about randomly in an attempt to achieve amenity cheaply. They generally look best when used as focal points, in clusters or groups, or in linear arrangements, for instance, along walls, fences or railings; or in some other coherent pattern that makes sense from the eye-level and high-level points from which the floorscape will be visible.

Concrete is a natural material for plant containers. It is robust, can be moulded into attractive forms and finishes, and does not require to be lined or coated as is the case with timber or metal. Also concrete is visually appropriate in the hard landscapes where plant containers are most frequently used.

Containers must be provided with drainage holes, and bottomless types are available to allow roots to grow down into the ground. All loose landscape furniture should stand on a hard base for stability, and if such hard surfaces are provided below and adjacent to plant containers, they should be laid to falls with drainage arrangements, to allow for hosing down. A guide to available types will be found in the COID publication *Street furniture from the design index.*

81 *Rectangular plant containers, of different sizes, constructed from standardised modular concrete forms.*

4.08 Bicycle parking

Stands should be provided near the points where people dismount, and should be immediately visible without being too visually or physically obtrusive. People on bicycles will not detour if they can avoid it. If there is nowhere handy for parking a bicycle when they dismount at the kerb, they will either leave the cycle at the kerb or start wheeling it towards their destination with the intention of propping it against a building or a fence. Stands must therefore be accurately positioned to intercept their routes.

Stands must be designed and located so as to look tidy, attractive, and visually at home, both when empty, **82**, and when full of bicycles—and, for that matter, when half-full. When there are no bicycles, the stand ought not to look naked. Placing stands where they are overshadowed by more dominant objects such as walls, trees or steps often helps to incorporate them into the townscape. There must be adequate circulation space round the stand even when it is full.

The first, and simplest, form of bicycle stand is the slot in a precast paving slab. This offers simplicity of installation as it is integral with the paving surface, with no vertical posts to interrupt the view, and it blends into pedestrian surfaces when not in use. It is important to ensure that the slot is drained. Another alternative is the combined use of tree grilles and bicycle stands. With slotted precast slabs, the slot is diagonal, whereas for tree grids the slot is parallel to the main axis. The diagonal slot can be supplied with the slot going either way, so that it can be approached from either side and will secure either the front or the back wheels of bicycles. The slab can be either proud of the surrounding surface or set flush with it. The decision should be taken early as slabs usually have a minimum thickness of 100 mm. Slabs are available with a slope on the top face. These slabs have a wide range of application when they are flush with the surface.

A second type of stand is the vertical support type, which restrains each side of the bicycle's wheel. This method requires a vertical fixing to a wall or bollard, or its own vertical section fixed into the ground finish or the pavings, and if handled badly it can clutter up a pedestrian area with lines of posts. If the vertical face to which the stands are fixed is a wall, the parking arrangement will, of course, be linear, but radial arrangements around bollards, flower bowls or trees can be very effective, **82**.

82 *Example of slot-type bicycle stands very attractively incorporated into brick and concrete paving (lecture halls at Cambridge).*

4.09 Trees and grass

Trees

If trees are incorporated in paved areas, a soil pocket is required round each trunk, and an adequate surface water supply should be able to pass through the soil to the tree's roots. If the earth round the tree is sealed, the tree may dry up and die for lack of water and air. Another important point is to leave enough space to allow for the tree's natural growth, including the expansion of the trunk.

There are three possible methods of detailing the tree surround. The first is a radial paving pattern, with the pavings draining towards the tree and watering the soil through a radial arrangement of gratings round the tree. In this way, the paving would continue right up to the central soil, the tree would be related to the whole pattern of the landscape and the surface water would be drained usefully away.

The second method is to lay a panel of setts round the tree, **83b**. The third is to lay manufactured tree grids as a collar. Manufactured tree grids can be obtained in cast iron, **83a**, or precast concrete. The precast concrete grids offer the possibility of a continuous concrete surface right up to the tree if used in conjunction with paving. The main support for the tree grids would be their outer edges. The soil pocket will allow for any irregularities in the shape of the tree and it is best to top the soil with 20 or 10 mm single-size gravel to restrain the soil and provide a tidier appearance. In this manner, the paving would be a continuous flat surface up to the tree trunk.

A kerb is sometimes misguidedly fitted around the tree to protect it and to edge the pavings. This should not be done. It breaks up the horizontal surface and forms a useless miniature flower bed for collecting litter, while offering only negligible protection to a hardy tree trunk.

Grass

If pavings are let into grass areas, or laid adjacent to lawns, the hard surface should always be at least 40 mm below the grass to allow for mowing. If grass is planted alongside a vertical hard surface, the edge should be kept 250 mm away from the vertical face, for the same reason.

83 a, b *Two forms of tree surround:* **a** *metal grille;* **b** *charcoal-grey split concrete bricks forming gently sloping surrounds.*

83a

83b

5 Play areas

5.01 Introduction

Play is a spontaneous activity, **84**, and planning (whether by educators, architects or landscapers) is likely to be unsuccessful if it attempts to impose, or if it assumes, a single pattern of use. Playgrounds should therefore encourage active, spontaneous and creative play. A conglomeration of complex gadgets for playing does not provide the necessary sense of challenge; half-finished components and materials to encourage active creative play are likely to be of more value, and materials for building and experimenting more popular, than formal games.

84 *Recreation area in Amsterdam, including sandpit and play shapes formed with precast concrete components.*

5.02 Play shapes

Play shapes such as boxes, houses, arches, castles and mazes, walkways, slopes, steps and ramps can all be created with an assembly of vertical and horizontal precast products such as pipes, kerbs, concrete blocks and paving, **85**. Square, rectangular, and circular pavings of various sizes can be used; bollards and plant containers have great potential; and ramparts, stepping stones, boundaries and terraces are easy to form with concrete components. Children find it easy to create their own games with random shapes such as the above, and concrete components are very suitable provided elementary safety precautions are taken. These include the absolute necessity of smoothing off all edges (easily done—see Information sheet 15) and taking care to avoid all possibility of children falling on to concrete from a height of more than 1 metre. Children must also not be exposed to the dangers of scrabbling down steep concrete slopes.

The design of play shapes must take into account not only the way in which the shapes will be used by children, but also ease of production and assembly and resistance to weathering. As far as production/assembly is concerned, the problems of transportation, handling, hoisting into position, and fitting together of large, heavy elements should be given attention.

When it comes to weathering, then material, form, surface texture and jointing are the four aspects to be taken into account.

● All *materials* must be colour-fast in sun, wind and rain; and if there is a possibility of colour-loss, the possibility of one material staining another (through rain run-off, for instance) should be considered. Any metals used must be resistant to corrosion from dampness or water, and inert in contact with the other materials specified (see Information sheet 17).

● The *form* of the structure should be so designed that it is self-cleaning. Concave surfaces in which water may be trapped should be avoided, or else drained through holes; because such collected water is liable to cause frost damage and spalling. True horizontal surfaces should also be avoided if possible, because they do not wash clean and dirt tends to collect on them. Streaking can be avoided by using drip mouldings to throw rainwater clear of vertical surfaces.

● With regard to *surface texture*, smooth finishes are not only more liable to cracking, crazing and staining than textured or porous ones, but make it harder to camouflage these blemishes. Staining may be due to the retention and drainage of dirt, to water flows across the surface which then carry the dirt down to form streaks or other patterns, or to chemical action and corrosion. Corrosive rainwater may drip from trees, from stone buildings, from parts of the play-shape at a higher level, or corrosion may be due simply to an industrial atmosphere. If staining is to be avoided, therefore, textured or porous surfaces combined with a shape in which water run-off is thrown clear or carried in well-defined channels is the best answer.

● Finally, *joints* require careful design in order to avoid water penetration by capillary action and erosion by rain in vertical joints with the possible danger of frost action. See Information sheet 8 for more detailed data on appearance and weathering.

85a

85 *Examples of play shapes and spaces created with precast concrete components:* **a** *playground in Düsseldorf. Unfortunately none of these examples exploit to the full the possibility of creating really complex, challenging ramparts to be climbed or mazes to be explored, but the potential is obvious;* **b** *simple arch made from standard precast concrete drainage pipe;* **c** *simple hide-and-seek.*

85b

85c

86a

86b

5.03 Sand, water and horizontal surfaces

Sand

Playing with sand challenges the imaginative and creative powers of the child and is therefore an important element in any children's playground, **86**.

All too often, a sandpit is too small. Children do not always want to sit still and make sand pies, but prefer to build castles, tunnels, mazes, trenches and canals. In contrast to the sandpit, a large piled-up mound of sand is not only more interesting to children than sand in a pit but has the advantage of being cheaper—a sandpit has to be cleaned and filled regularly with new sand. The greatest variety for play can be achieved by a combination of sand-heap and sandpit, ie, a sandpit which is generously filled with heaped sand at intervals.

86 *Sand play areas:*
a *rather larger, more complex concrete sculpture/sandpit combination;*
b *concrete Gulliver reclining*
in sandpit in London housing estate;
c *simple, small (probably too small) sandpit in Berne housing estate.*

86c

87a

87b

Water

In addition to its enormous potential for adding interest to the townscape, **87**, water is as popular with children as sand, **88**, **90**. In particular, they are delighted when they can play with sand and water at the same time. In playgrounds where bathing facilities for small children are unsuitable, a drinking fountain may even be sufficient; or perhaps a simple concrete trough, a hollowed-out concrete block or a tree trunk.

Two types of paddling pools are quite widely used: the pool of even depth all over, with approximately 200 mm of water, and the pool with a floor which slopes gradually towards the middle, leaving the edges shallow. Often a paddling pool can be combined with fountains or showers, **89**. Paddling fountains instead of paddling pools can be built with painted concrete pipes of about 2 m in diameter and of different heights. Water can be supplied from a jet in the highest basin and allowed to run over the periphery of one basin to the next and down to the lowest. If the water level in the individual basins is only 100 mm deep it would not be dangerous even for the smallest child. These paddling fountains have several advantages compared with paddling pools: the children are divided up into small groups by the different basins and this is more appropriate to their ages than a noisy pool which is meant to accommodate everybody. Two or three children have their own 'lake'. The different heights of the basins usefully encourage movement from one basin to the other. In addition, the paddling fountain, because of its relatively shallow water level and the smallness of the basins, is less interesting to older children so that the smaller ones can play unobstructed. The paddling fountain has a further advantage that, as a result of the constant flow of water, it does not become so easily polluted and therefore only requires cleaning weekly in contrast to the paddling pool, where daily cleaning is necessary. A combination of paddling pool and paddling fountain can also be successful: it consists of a fair-sized paddling pool with a paddling fountain and several basins in one corner. In this way, it is possible to combine the entire paddling equipment for small children and those of school age, without disturbing or endangering the former in the fountain basins. In cooler weather, the pool can be left empty and only the paddling fountains filled with water.

87 *Combination of concrete and moving water can make very effective townscape:*
a *a water garden in central Portland, Oregon, US. Rapids and waterfalls flow over man-made rocks and ramparts; and stepping stones allow pedestrians to cross the water area (see also* **40***);*
b *water garden fountain with dancing water walls when fountains are turned on, pure concrete sculpture when dry;*
c *concrete water wall in Vienna park.*

87c

88

89

Horizontal surfaces

Playgrounds with natural or artificial hills provide far more interest than a monotonous flat area. A slide should not be placed unimaginatively in the centre of an open space when it could well utilise a slope and serve as an entrance to the playground. A sandpit need not be a square box but rather a free form which exploits the plastic nature of concrete.

There are two forms of surface which hold fascination and can be extremely effective if handled well. The one is water and the other is rocks. Both materials are equally at home in the pedestrian environment and in play areas.

A rockery can provide a dramatic visual focus; and by exploiting varying levels and by directing pedestrians either around or partially through the rockery to reach a sheltered area, an increased impact can be achieved.

The second use of rocks is not so frequently seen, and may be more suitable for exposed urban areas. The principle is to create a rock-strewn surface which is impassable and where the centre is inaccessible. Solidity and ruggedness are characteristics of this feature. The boulders and rocks themselves should be in random sizes up to 2 m across, and 'scattered' at random over a very coarse loose gravel bed, which should be 150 to 200 mm deep. Evergreen shrubs can be planted between the boulders and the gravel raked into mounds and hollows. The rocks and gravel need to be chosen carefully, but local stone is often suitable, both for aesthetic and economical reasons.

88 *concrete play sculpture in Letten swimming pool, Zurich;* **89** *combination of fountains, shallow pools and concrete forms in Stuttgart promenade;* **90** *simple bowl fountain intended mainly as a visual focus, but obviously enjoyed by the children.*

6 The vehicle environment

6.01 Introduction

In the vehicle environment, as in the pedestrian, it is helpful to the designer to think in terms of two kinds of space: *paths*, **92a** (spaces which cater for movement and travel) and *places*, **92b** (spaces where movement comes to a stop). As explained in section 1.02 this is not a watertight division and it should certainly not be pursued to absurd lengths, but it does provide a useful framework for the presentation of information. The vehicle environment will therefore be dealt with, here, in two sections: movement and parking.

92a *Paths cater for movement; and* **b***, places for vehicle parking. Latter example shows parking place for* *fire-fighting equipment, with profiled finish (Mono BG slabs) to deter pedestrian access.*

92a

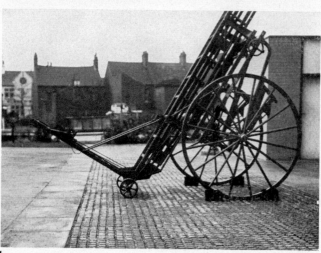

92b

← **91** *(community centre in Esslingen, Germany) shows sunken garden forming buffer zone between vehicle and pedestrian access and community centre.*

6.02 Movement

It is probable that some highway and motorway-building will continue for the foreseeable future (even if at a reduced scale), and the urgent issue for landscapers is how to improve their design. The examples shown in this section demonstrate that when they are constructed in sympathy with the local landscape they can be visually successful; and provided they are built as part of a balanced private/public transportation plan, with due regard to noise and air pollution problems, and social needs, they can make a positive contribution to the built environment, **93**. There is no reason in principle why urban highways cannot be used as positive design elements in reshaping the form of the city, as long as planners accept and solve the social and economic problems created by their construction.

Ideally, motorways should achieve a high operational standard and at the same time fit into a built-up area with a minimum of disturbance to people and property, and without impairing amenity, **94**.

Unfortunately it is often impossible to reconcile all these requirements within the existing cost limits.

A motorway which follows the profile of the existing ground is invariably cheaper than one which passes over or under obstructions, but in built-up areas a ground-level motorway is often impossible unless all the existing roads in the area run parallel to the motorway, thus obviating the need for crossings. The latter situation seldom exists, therefore the choice is generally to build the motorway either above or below ground.

Sunken motorways

From an amenity point of view, the sunken highway, **95** is preferable to the elevated one; but it may be ruled out by the amount of dislocation it causes to existing underground mains and services.

93 *Clean, simple lines of the elevated motorway in Stuttgart form an attractive element in the townscape.*
94 *Planting along the retaining walls provides welcome relief and contrast to the dominant parallel hard surfaces of the sunken motorway.*
95 *The new town of Leverkusen is based on an extensive network of motorways which have been depressed to facilitate integration into the new urban fabric.*

93

There are three fundamental types of sunken motorway construction—an open cutting; cut and cover; and a tunnel. Where a *cutting* is driven through an area and the motorway constructed at the lower level, noise nuisance will be substantially abated, 95; but the whole of the land occupied by the cutting has to be acquired and any property in the way has to be bought and demolished. Bridges to carry road crossings must be built across the cutting and underground services within the route of the motorway need to be diverted. To reduce the land-take involved and the amount of demolition required, the width can be reduced by constructing vertical retaining walls to support the adjoining land instead of forming slopes. The former tend, however, to be very expensive structures and their cost must be balanced against the value of land and property saved. Where land values are very high the motorway could be roofed over and the site utilised for parking, offices or industrial building, with land reserved for mass transportation and employment centres.

Cut-and-cover is the term used to describe building a road in a cutting with vertical retained sides, which in turn support a cover, 97. A covered motorway will abate noise nuisance even more than the open cutting, probably eliminating it entirely. This form of construction can be designed to permit landscaping or indeed redevelopment three to four storeys high. An alternative is to provide a lighter cover and design the building to span the gap. Where natural ventilation is used, it is important to design the surface openings so that they cause the least visual distress and do not diminish the noise reduction advantages that would otherwise be obtained. Where artificial ventilation is used there is an advantage in providing a central wall in the tunnel to permit the use of a simple longitudinal ventilation system in which fans, situated at each entrance, force in air to aid the natural air-movement caused by traffic moving in one direction.

96

The *bored tunnel* (or series of bored tunnels) will eliminate many of the problems associated with both the open cutting and the cut-and-cover road. Existing buildings need not be demolished or disturbed, and it may be possible to avoid much of the disturbance to underground mains and services caused by the other two methods. But the deep tunnel is extremely costly and with a tunnel of any length, lighting and ventilation must be provided. A further problem is the difficulty of forming connections to the tunnel.

Elevated motorways

An elevated motorway can be constructed either by means of an embankment, or flyovers, 96. The land occupied by the embankment type is sterilised for all time and the completed motorway may form an objectionable barrier between the severed areas. Flyovers do not have these particular disadvantages, but they tend to cost much more.

Interchanges for elevated roads occupy very large areas of land and, while they can be fitted into the countryside without difficulty, they tend to disrupt the urban scale and grain disastrously, 98. The number of locations where they can be accepted in planning terms are therefore limited. An additional problem associated with them is the noise and fume nuisance generated in their immediate vicinity by the dense streams of accelerating and braking traffic, much of it consisting of heavy trucks.

96 The use of large-scale structures set in soft landscape can be an attractive feature in the urban fabric.

97

97 *Urban motorway in Berlin which features cut-and-cover technique.* 98 *Large-scale motorway structures such as that in Dusseldorf occupy large tracts of land and tend to isolate the surroundings from pedestrian use.*

98

Noise

Traffic noise is a major problem in the integration of motorways in an urban area. However, continental experience indicates that good planning, the use of acoustic barriers, **102** (and perhaps in future the reduction of noise at source, by legislation and technical progress in vehicle design), can reduce the problem considerably. Where the design of the road provides uninterrupted flow with easy gradients (particularly desirable where heavy commercial vehicles are concerned), the noise levels are much lower than those associated with roads involving steep gradients, and braking and accelerating. Obviously, therefore, the noise problem must be considered from the earliest stages in the investigation and location of routes, particularly where they pass through residential and recreational areas.

The best results in open cuttings are obtained by having vertical walls, preferably with a sound baffling finish such as open-textured cored lightweight blocks, left exposed. If the cutting is shallower, upstand walls at ground level will assist in noise reduction—the higher they are the better*. Through areas of high amenity the cutting can be covered in and this can make the noise problem minimal.

The visual impact of acoustic barriers must be given careful thought, **101**. In the case of high rise blocks near depressed or grade roads, even allowing for the reduction by distance the only solution may be double glazing and in extreme cases air conditioning. Sealed double glazing with plate glass can have a sound reduction of 35 to 40 dBA (provided of course windows

*A method for calculating appropriate wall height is given in BRS Digest 135 Nov. 1971) *Motorway noise and dwellings.*

99

are shut, which means that natural ventilation may be ruled out). A drastic solution is to change the land use adjacent to the motorway to non-noise sensitive use, but this is unlikely to be an economical proposition in most urban areas. A more practical method is to adapt the redevelopment to the presence of the road, orientating residential blocks with either blank walls or rooms with non-opening windows on the motorway side. Even where the motorway is elevated, a four-storey terrace block planned in this way, parallel to the road, can act as a shield for housing further away.

Design considerations

Although it is important that urban motorways do not spoil the areas through which they pass, it is even more important that they make a positive contribution with their potential value exploited as a new element of design.

One of the more difficult design problems in urban areas is the composition of views seen at speed. Both safety and pleasure depend upon the road unfolding itself inevitably before the driver with smoothness of line and contour without unexpected checks and jerks, **99**.

Bearing in mind the functional requirements of the motorway and the need to integrate it with the urban environment the standards of geometric design should be of the highest order, **96**.

The selection of the alignment should take full account of the need to permit junctions to be sited so as to give the maximum

99 *Simple clean lines of the motorway interchange*   *outside Cologne facilitates use and minimises driver error.*

100

101

102

operational service compatible with planning considerations.
In general, motorways should have gentle curvature which present an ever-changing backcloth and provide visual stimulation for the motorist. Horizontal and vertical curves should be as long as possible with no short curves or straights. Where two curves are both visible to the motorist they should not be connected by short straights but should flow one into the other. If two curves follow each other in the same direction, they should be connected by a flat curve to provide a flowing alignment. The best roads have an appearance of inevitability which reflects both the smooth, flowing line needed for safe and pleasant driving, and a feeling for the surrounding townscape, **2, 24**.

The landscaping of motorways is all too often looked upon as the planting of trees after the motorway has been completed. But no amount of tree-planting can disguise a poor design; and the true function of tree planting must be to help merge a sensitive design into its surroundings, **100**.

Surface treatments should be chosen with special care when they will be seen at close range by pedestrians, so that they will look pleasing not only when seen from a distance but also close up.

Finally, the plasticity of concrete enables profiled finishes (with deep relief, sculptural forms, and geometric patterns) to be used, exploiting the effects of light and shadow.

As a general rule, the more textured and profiled the concrete surface, the better it will weather, as minor blemishes and defects will be imperceptible. Such textured and sculptured surfaces can be produced at a price that is highly competitive with those of traditional materials, by using plastic linings as moulds. They provide almost complete pattern flexibility, can be used for non-standard shapes, for highly complex shapes unsuitable for steel or timber moulds, and will produce a concrete surface that is smooth and uniform in colour, with an eggshell finish.

100 *Planting incorporated as an essential element in design of urban motorway in Essen.*
101 *The walls to the underpass at Cologne University have been designed not only to reduce* traffic noise but also to encourage plant growth.
102 *Retaining walls to depressed motorways can be designed to provide acoustic baffles and thus reduce noise at source.*

6.03 Parking

There are four basic kinds of parking solution—ground level parking, **103**, parking structures above ground, **105**, underground garages, **106**, and garages combined with other functions, **104**.

Ground level parking is often less expensive to provide than the other solutions, but space availability is such a problem that it can seldom be the whole answer in modern cities. Underground garages are expensive but have enormous environmental advantages, being able to provide parking facilities underneath parks and squares, and without disrupting the cityscape, **106**. For above-ground structures a distinctive architecture of parking has emerged, **105**; but it would be idle to deny that this has often been visually appalling. The scale of such a structure in relation to the surrounding buildings, and the problem of incorporating the massive spiral ramps into a traditional street facade consisting usually of small-scale horizontal and vertical elements, has defeated many architects; and the multi-storey car park seldom looks at home in the traditional townscape. There is however no fundamental reason why this should be so, as some of the better examples indicate.

Design considerations

To attract the car driver, a carpark needs:
● To be within about 400 metres (five minutes' walk) of the destination—shops, station, or entertainment, **103**.
● To provide a clean, dry, even surface for walking on as well as driving.
● To have easy entrance and exit facilities, both for cars and pedestrians.
● To have low, or no parking charges.
● To be well signposted and have parking spaces clearly indicated within it, **103**.
● To be limited in size to a maximum of about 400 cars. Unless rapid dispersal from the exit areas can be guaranteed, the nearby roads will become congested at peak periods.
● To have—and this is most important—low maintenance costs, with a floor surface that is not affected by heat, rain or oil drips.

A space intended to park cars has to serve two functions—one with and the other without cars, **107**. A large expanse of monolithic carpark in a grey, black or tinted surface, marked out into rectangles in maintenance-requiring white paint, is nearly always unattractive. If bays are required these can be

103 *Well defined car parking both clearly marked and in close proximity to shops.*
104 *The careful integration of parking along the river minimises what would otherwise cause considerable disruption (Lyon).*
105 *Well defined statement of multi-storey car parks in Cambridge and Shrewsbury.*

104

105a

105b

106

used to add scale. The difference between parking areas and circulation areas can also be used to advantage, **108**. For example insitu concrete can be used for service roads and circulation areas, with precast paving for the parking bays. To alleviate the impact of serried rows of metal, planting in the form of both shrubs and trees offers many advantages. The planting needs to match the surroundings but the inclusion of some evergreens will be a benefit.

Large outdoor car parks benefit from strips or clumps of planting, and this is now a generally accepted practice. The possible nuisance of leaves is more than outweighed by the very real benefit of both screening and shade. Species with smaller leaves, or with leaves which decompose relatively easily, or with needles, cause least trouble and shrubs can be also very effective. Trees that should not be used in carparks because they drop gums, berries or heavy leaves are: lime varieties (italia platyphyllos, tilia tubra, tilia euchluva) maiden hair (ginto biloba) and horse chesnuts (acer aesculus, hyppo-castorum). If a few trees or even shrubs exist, they should be kept for at least a year, to see if they are not in fact an asset rather than a liability. Only species sympathetic to the regional landscape should be introduced; otherwise the screening itself becomes an intrusion.

Where planting is included in a carpark it is advisable to raise the soil beds. This reduces the likelihood of damage and increases the effect of the screening. Dwarf walls of concrete blocks suitably matched to their surroundings can be used (see Information sheet 5). Split blocks offer a wider range of aggregate colours and textures and tend to be more in keeping with the cars and with any concrete used in the carpark. The regularity of the car spaces between the trees may themselves force a regular pattern upon the planting of the trees. This must be watched if an informal effect is required. Deciduous planting provides less screening in winter than in summer. This is why some evergreens are an advantage. Trees may give shade to cars in the summer, and their shadows may help to disguise the shape of the cars.

The use of coloured or textured paving surfaces can be a great advantage, both for functional and aesthetic reasons.

Bearing these considerations in mind it follows that concrete is one of the most suitable materials for paving a parking area that will be both economical and good-looking. As is well known, maintenance costs for a concrete paved surface are practically nil; the surface is not affected by oil droppings, is not softened by heat, and does not collect the dirt. Concrete paving can be constructed satisfactorily on ground which is

107

108

107 *Mono BG slabs used for parking area to provide surface which is attractive both with and without cars.* **108** *The use of different coloured precast paving to define different areas and articulate parking space.*

poor in bearing quality; so concrete carparks can be sited in areas for which it would be difficult to find other uses.

Moreover, construction of a concrete carpark is not a disruptive activity. It can be carried out with relatively small plant, and the use of ready-mixed concrete still further reduces any site work involved (see Information sheet 2).

Layout

A parking area may be considered, broadly, as two units: the parking bays proper, and the roads linking them. The parking bays should be clearly marked, to enable the driver to park without confusion or doubt, the roads should be wide enough to enable him to do so without risk of hitting other parked vehicles. Most cars in Europe can be accommodated in an area 5 m by 2·5 m still leaving sufficient space for the opening of doors and 'errors of judgment'. It should be remembered, however, that some cars are over 6 m long, and as much as 2·5 m wide, so some extra long parking spaces may have to be provided. If possible a bay size of 6 m by 3 m should be adopted as the standard though, if space is very limited, this could be reduced to 5 m by 2 m. The different parking bays for different vehicles—normal cars, extra large cars and motor-cycles—can be indicated by different coloured concrete paving.

Possible parking arrangements are:
● parking along the line of the 'road' as in kerbside parking;
● parking at 90° to the line of the access road;
● parking at an oblique angle generally 45° to the line of the access road.

Of these alternatives the first can in most circumstances be disregarded as impractical and uneconomic in space. Parking at 90° requires a minimum road width of 7 m; with parking at 45° the road width can be reduced to 6 m, and, if one-way traffic is used, can be as little as 3 m. Typical layouts using 90° and 45° parking are shown in the *AJ Metric Handbook*.

Surface finishes

Choice of material depends on the nature of the vehicular and pedestrian traffic, natural ground conditions and the character of the area. The types of surface suitable for carparks in urban areas can be either insitu or precast concrete construction.
● Hollow paving in grass which is suitable for heavy occasional use and lighter everyday use, **109**. Hollow concrete paving blocks through which the grass grows provide a

'reinforced field', a most effective answer which has been used with success particularly in Denmark and Germany and is now available in the UK.

● In situ concrete construction including no-fines concrete either totally or in combination with:

● Precast paving. Although a high standard of riding quality is not necessary on a carpark, generally a smooth surface is desirable, both for drainage, and for ease of walking and cleaning, **103**. For this reason a good standard of finish should be achieved. A tamped finish, unless it is fairly fine, is not generally good for pedestrians, nor does it assist surface cleanliness; the 'sandpaper texture' formed either with a float, or with a light brush is better.

A brushed finish is attractive, cheap and easy to obtain. The degree of roughness will vary with the stiffness of bristle, but as pedestrians have to walk over the carpark a very rough texture should be avoided.

It is sometimes desirable to mark off certain areas of the park with a contrasting colour, and with concrete paving this can be easily done, **110**. Different colours can be provided by using coloured cement—the use of the deeper colours such as red, buff and khaki does not involve any significant additional cost. In the comparatively thin slabs used in carparks it is generally more economical to use coloured cement for the whole depth of the slab, rather than to limit it to the top 50 mm, as is sometimes done in the case of thicker slabs. Besides its practical use, the introduction of this sort of colour variation can greatly enhance the appearance of such areas. The same applies to variations in textures. Exposing the aggregate —either by watering and brushing off the surface mortar, or by spreading on the surface a layer of a specially chosen aggregate—can be an effective and economical way of varying the surface texture (see Information sheets 9-13).

If it is decided to apply a surface layer of aggregate it should preferably be of 20 mm or 25 mm single size, and be forced into the plastic concrete with a vibrating beam. Surface mortar is then removed by washing and brushing.

Methods of producing exposed aggregate and textured finishes are described in Information sheets 9-13.

Surface water drainage

The fact that concrete paving can be laid to shallow falls, combined with the knowledge that the slabs will not deform under traffic, considerably reduces surface water drainage problems.

It is recommended that a minimum crossfall of 1 in 60 is adopted and that longitudinal falls should be not less than 1 in 200. Small carparks on reasonably level sites can be conveniently drained by falling to channels at the sides of the area, and then to gulleys. This system reduces the number of drainage pipes beneath the pavement. In larger areas the most common method is to construct the slabs so that they fall to collecting gulleys within the area.

Some authorities use open drainage channels in order to reduce the number of underground drainage pipes. These channels can be constructed insitu. In this case a gap of not less than 300 mm is left between the slabs, and the channels constructed after the slabs are laid, using the same mix as that in the slab. The simplest technique is to tamp the channels longitudinally, working off suitably shaped timber and forms, and using the alternative bay method of construction. The channels should be constructed in lengths not exceeding 2 m and the thickness of concrete beneath the valley of the channel should be not less than 100 mm. The joint between channel sections should be a simple unsealed butt joint. Precast channel blocks can also be used, in which case the joint between the blocks and the adjoining slabs should be at least 10 mm wide and sealed with a sealing compound. The joints between blocks should be made with mortar spread over the whole of the ends of the units. An expansion joint should be

109

110

109 *Hollow concrete pots used at the Sunderland Civic Centre permits intermittent access for service and fire vehicles.*

110 *Different coloured precast paving is used for delineation of space.*

made at intervals of not less than 15 m in each channel length, by placing a 10 mm thick layer of compressible filler material between the blocks. These expansion joints must be sealed and should coincide with transverse joints in the pavements.

Junction of carparks with an existing road

At the junction of a concrete carpark and a bituminous surfaced road a row of granite setts should be laid at the end of the slab. The setts should be at least 100 mm deep and supported on a 150 mm by 150 mm concrete foundation. They should be bedded on a 3:1 sand/cement mortar and grouted up with a 2:1 sand/cement mortar.

When the carpark adjoins a concrete road no problems arise—a butt joint with a sealing groove is all that is required.

Information sheet 1
Pavings (precast)

1 Use of precast slabs

1.01 Introduction

Precast concrete paving is used for areas where durable, self-cleaning, non-slip surfaces are required. It is eminently suitable as paving in urban areas where these qualities are particularly desirable and where, if required, coloured and textured flags can provide an aesthetically pleasing effect. To achieve these objectives it is essential that well-made paving flags are properly laid and the purpose of this information sheet is to describe the materials and the simple construction techniques that should be adopted to ensure that the paving has a long life and a pleasing appearance.

1.02 Where to use

Well-laid precast slabs are usually superior to in situ alternatives such as concrete or tarmac both in appearance and in wearing qualities. But unfortunately they are usually also more expensive in first cost (though not necessarily in the long term—see next paragraph). It is therefore best to restrict the use of precast pavings to limited areas, where their attractive appearance and durability are most desired.

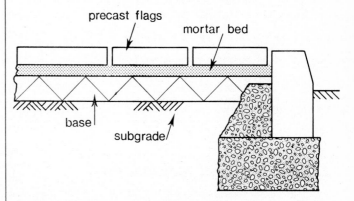

1 *Typical section through a paved footway.*

2 *Laying a precast paved pathway in soft landscape.*

An obvious application is the formation of a durable and attractive pathway for fairly intensive traffic, using precast slabs, leading through an area with a cheaper and probably less durable overall finish, 2. This kind of use is sensible both in terms of economy and of rational landscape formation (with primary paths clearly and visibly differentiated from the rest of the landscape, by means of surface finish—see for instance 10, 20 and 29).

1.03 Comparative costs

The following figures are not intended to be precise, but merely to give designers a very broad indication of the order of difference, in first cost, between various alternative finishes (including bases). If the cost of precast slabs is taken as 100, gravel would be approximately 35; 60 mm tarmac 70; 100 mm in situ concrete 75; cobbles set in concrete 240; brick paving 250 to 375; York stone 550; granite setts 650.

However, it must be remembered, in looking at the above figures, that while a bituminous surface, for instance, will require resurfacing in five to 10 years, a good flagged path will last 25 to 30 years with little need for maintenance, and in addition provide a surface of the highest quality, if the elementary precautions given in 5.01 and 5.03 are followed.

2 Slab types

The range of precast concrete paving products is very wide and standard sizes are given in Table 1. These notes deal with types of precast concrete paving commonly supplied in the UK and abroad. They outline the principal methods of manufacture and comment on the quality and type of finish which can be obtained by the various processes.

Table 1 Standard sizes for precast concrete slabs

Type A	600 mm	× 450 mm
Type B	600 mm	× 600 mm
Type C	600 mm	× 750 mm
Type D	600 mm	× 900 mm

In all cases, thickness 50 mm for pedestrian use; 63 mm for occasional vehicular traffic.

2.01 Methods of manufacture

Hydraulically pressed paving, although widely available in the UK is rarely encountered in Europe. The process is based on a wet concrete mix from which the surplus water is extracted by top pressure and filtered by a paper membrane.
Vacuum pressed as the name indicates employs the vacuum concrete principle involving compaction of a wet mix under vibration. Surplus water is extracted by a vacuum pump.
Dry-pressed-vibrated paving is manufactured from an 'earth-dry' concrete mix subjected to combined pressure and vibration. This process is widely used on the Continent and is becoming more common in UK.
Dry-pressed-tamped products are formed from an 'earth-dry' mix compacted by mechanical tamper action. This process is used mainly on the Continent but also in Britain.
All the above processes are carried out on machinery capable of high output. The paving produced has a low ultimate water/cement ratio, ensuring a rapid gain of strength and subsequent durability. The production process usually includes sophisticated handling equipment enabling large

3 Hydraulically pressed slab with exposed aggregate surface.
4, 5 Exposing the surface by spraying and brushing.

6 Manufacturing concrete paving slabs by open mould process.

3

4

5

6

throughput with a minimum of labour.
Vibrated slabs are manufactured in individual moulds compacted on vibrating tables similar to cladding and other types of precast products. Demoulding usually takes place after 24 hours.

2.02 Quality

All paving slabs should comply with BS 368 which lays down tests designed to ensure adequate strength, wearing properties and resistance to frost attack. Manufacturers of precast paving normally carry out tests to verify their standard of quality control. All the processes listed can manufacture products capable of meeting this standard. Frost attack, in cases where de-icing salts are employed, provide the most rigorous conditions of use, and slabs manufactured by any process can fail to perform if their initial quality is deficient. Reputable manufacturers employ quality testing procedures and can indicate pavings which have been in use without any deterioration after many years of service in widely varying conditions.

2.03 Appearance

The appearance of paving is determined by the methods used in manufacture. Some of the more common finishes are listed below.
Hydraulically-pressed—normal finish which has a smooth surface with 'pimple' imprint caused by the perforations in the press head.
Hydraulically-pressed—other finishes in which the surface is smooth but with various patterns obtained by using other types of mesh etc during the pressing process.
Vacuum pressed products are generally similar to hydraulically pressed in appearance although a rough textured finish is also available.
Dry-pressed paving whether vibrated or tamped tend to exhibit more of an 'open' texture similar to other forms of dry cast concrete. Surface finishes are usually applied to top layer only and these can be smooth, textured or with exposed aggregate.
Vibrated slabs are similar to normal vibrated concrete and can be produced with virtually any desired surface finish. Patterns or special texture finishes can be obtained by casting against the mould, face-down. Rough textured or aggregate-faced surfaces are usually applied to the upper surface.
Capital-intensive processes such as hydraulically-pressed plants are largely designed to execute long runs of standardised products. The introduction of special shapes and sizes is costly and only justified if very large quantities are required. Most presses in all categories set limits to the overall size of the slab available. Thicknesses can be more easily varied if required. Smaller presses and vibrated processes are more flexible, and special requirements are relatively easier to accommodate.
Most processes cannot manufacture slabs with steel reinforcement. On occasions where slabs have to support heavier loads than those covered by standard paving it is generally preferable to provide the necessary load bearing strength using a specially prepared concrete base.

2.04 Smaller units

Smaller units of concrete paving similar to tiles and setts, are also widely available and are frequently used where demarcation is required for some special purpose—parking areas for instance, or for borders or surrounds, 44. Cobbles set in concrete, 7, can also be used for special areas such as hazards, 28, to keep people away from grass and plants, 67, and at the same time they can provide a very attractive contrast with larger slabs, 90.

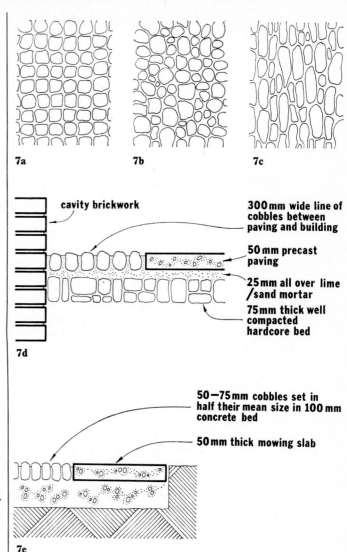

7 *Cobbles set in concrete. Cobble patterns include coursed, a; random b; and flat parallel, c. Typical cross* *sections shown in d and e: cobbles are bedded in 1:2:4 mix concrete, of thickness depending on cobble size.*

3 Design considerations
3.01 Bonding patterns
8 and 9 illustrate some of the bonding patterns which are possible, using standard flags.

3.02 Joint design
Pavings can be laid with open or closed (ie 'tight') joints. Open joints are classified as those wider than the 5 mm joint normally used (say 10-15 mm), but if the joint becomes too wide the mortar will crack away from the pavings. The mortar should be 1:3 cement/sand. The joint can be raked out to whatever level is desired, but water can collect in raked joints and cause cracking due to freezing. The sand used to finish the joint can be changed from that used for the joint itself, to contrast with the colour of the paving slabs. Coloured cements can also be used but great care should be taken not to mark the paving with the mortar. Concrete pavings need not suffer from excessive thermal or moisture movement, provided a few precautions are taken. For instance, if long lines of butt-jointed pavings are envisaged then some of the joints should be left a little wider than the normal 5 mm or so, to allow for movement. Normally, however, the joints are not tight enough to cause trouble, and the base should be stable enough to avoid any buckling due to movement.

3.03 Pavings abutting buildings

Where a paved area abuts a building, a harsh joint line can develop. It may be beneficial to use a different material or texture in the pavings for 300 mm or 600 mm along the base of the building. A textured or aggregate paving, cobbles, **7d**, setts or gravel would be suitable, **31**.

4 Handling and stacking

4.01 General principles

Although precast slabs are heavy and dense, they are nevertheless brittle, and it is important that edges and arrises are protected from chipping and breakage. The following guidance applies far more to 'faced' decorative pavings than to ordinary flags, but it would be good practice if it were applied throughout.

It is extremely important to provide ample space on site for storage and stacking and to arrange slabs so that they are easily accessible when required. They should never be stored where there may be drips from roofs or canvas covers, and when they are not being used their surface should always be covered with non-staining waterproof paper or polythene sheet.

Slabs must never be walked on before they are laid, or be handled with dirty or greasy hands, or with dirty ropes or cables. Nor should wet slabs come into contact with dirty sacks or pieces of timber, which might discolour them. Hydraulically pressed slabs smaller than 900 mm in width will resist most handling stresses without breakage, but care should be taken with larger panel sizes. Slabs should always be set down very carefully during handling, **11**, and hoisting cables, if used, should not be allowed to come into contact with the face of slabs. The use of pinch-bars or cross-bars for moving slabs should be avoided: if a bar must be used, a pad should be placed between it and the slab.

5 Laying

5.01 Precautions

To obtain a trouble-free paved surface which will last for many years without surface defects or maintenance, the following elementary precautions should be taken.

● The subgrade must be uniform and well compacted in order to prevent differential settlement (if the soil is cohesive, rolling will suffice; if it isn't, cement must be added to it as described later). Give particular attention to trench backfilling.

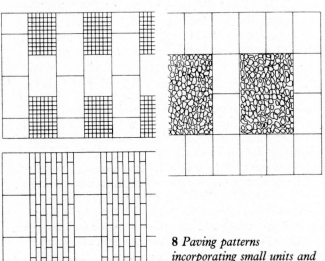

8 *Paving patterns incorporating small units and cobbles.*

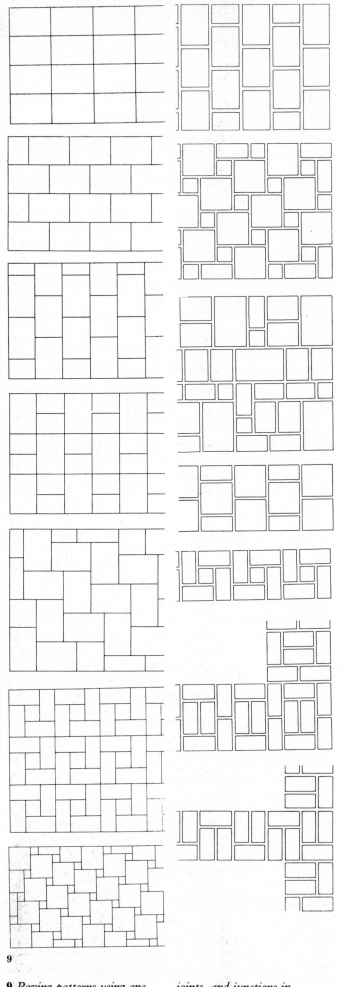

9 *Paving patterns using one, two, three and four sizes of slab, showing bonding patterns, open and closed joints, and junctions in pathways.*

● Materials used for bases must be stable and immune from the effects of water and frost, and well compacted to a uniform density. Loose, coarsely graded clinker, or fine underburnt ash cannot be compacted properly and should not be used. Stabilisation of base materials with cement should be considered when poor materials have to be used.
● As the potential life of flagged paving is 25 to 30 years, it is well worth spending extra money on good laying; and for this reason it is suggested that the whole-bed method be used (see paragraph 5.02).
● Joints should be carefully hand-pointed in 1:3 cement/sand mortar, care being taken to ensure that they are well filled. Alternatively, dry mortar should be swept into joints and allowed to hydrate naturally.
● At all footpath crossings, at corners, and wherever vehicles are likely to run on the pavings, the concrete bed should be at least 100 mm thick.
● In situations where pedestrian traffic is heavy and where goods are unloaded across the footpaths (eg shopping areas), flags should be at least 63 mm thick.
● Concrete edgings should be provided to all edges to prevent ingress of water to the base.

5.02 Methods of laying

There appears to be no generally preferred method of laying precast concrete flags, and some methods in common use are given below. The whole-bed method may however be the safest.
● Whole-bed method: A 25 mm thick bed of wet-mix lime and sand mortar (proportions 1:8 to 1:10) is laid under the whole slab, **10**. Joints between slabs are slurried in wet with the same mix, or alternatively (and perhaps preferably) a mix of dry cement and sand is brushed into the joints. As a third alternative, neat cement may be swept into the joints on still days.
● Dry method: Flags are laid on a dry mix of sand and cement or lime, **13a** to **d**, with tight joints which are slurried in with wet 1:3 cement/sand mortar. This method is not popular in shopping areas because of the inconvenience caused to pedestrians during and after the slurrying-in process; and sometimes a dry mix of cement and sand is swept into the joints instead.
● Five-spot method: The flags are laid on five spots of mortar, one spot at each corner and one in the centre of the flag. The spots are preferably not less than 150 mm in

12

diameter and not more than 25 mm thick. Hydraulic lime and sand mortar is used in mixes varying from 1:3 to 1:8. The flags are tight-jointed and the joints slurried in with the same mix. Sometimes a cement/sand mortar, in mixes varying from 1:2 to 1:12, is preferred for the slurry. The practice prevailing in some areas is to butter with cement mortar all sides of the slabs, fit them tightly together, and finish the joints by pointing. This method was probably used in the days of natural stone flagging when the sides of the slabs were not so true as those of the present-day precast slab.

5.03 Preparation of the subgrade

A properly prepared and compacted subgrade that provides uniform support to the construction is essential for the satisfactory long term performance of a paved area. The subgrade should be prepared by first removing any topsoil containing organic matter followed by any further excavation and filling so as to achieve the required formation levels to within a tolerance of ±25 mm of the true levels.
Filling material should be spread and compacted in layers not exceeding 150 mm compacted thickness. If the natural soil which has been excavated is unsuitable as fill it should be replaced by suitable granular, hoggin or similar stable material.
The whole of the formation including any fill areas should be fully compacted with at least a 350 kg vibrating roller, **12**, or a suitable smooth wheeled roller and where necessary a weedkiller used in accordance with the manufacturer's instructions, should be applied to the formation. It is important that the backfilling to trenches is fully compacted to prevent future settlement and to provide support to the paving similar to the remainder of the formation.
Excavated material should be protected from the weather by waterproof sheeting to retain the natural moisture content and it should be re-compacted using power rammers or small vibrating rollers in layers not exceeding 225 mm loose thickness. Alternatively, if the excavated material is unsuitable for re-use, the trench should be backfilled with stable material similar to that used for filling. Flooding the

10

11

10 *Whole bed method where a 25 mm thick bed of wet-mix mortar is laid under the slab.*

11 *Placing exposed aggregate slab in position.*
12 *Vibrating roller.*

trench with water will not achieve full compaction of the backfilled material and should not be permitted.

Edging strips. Where the paving abuts areas of vegetation or where it is possible that water might enter under the paving, precast concrete edging units complying with BS 340:1963 *Specification for pre-cast concrete kerbs, channels, edgings and quadrants* and laid on a concrete foundation should be provided along the edge of the paved area.

Base. The function of this construction layer is to provide adequate and uniform support to the precast paving. Several types of material are available for this purpose and hoggin, well graded crushed stone or gravel, cement-stabilised materials, well-burnt clinker or hard colliery shale are suitable. The compacted thickness of the base should not be less than 75 mm and the layer should be fully compacted by rolling with a 350 kg vibrating roller or a suitable smooth wheeler roller to produce a dense, even surface to within ±12 mm of the correct levels. If a dense, closely-knit surface is not achieved then additional fine blinding material should be rolled in as necessary. Where the subgrade consists of suitable granular material a sub-base layer will not be required but care should be taken to ensure that the surface is fully compacted by rolling and that any open areas in the surface are filled with fine blinding material.

5.04 Defects

Undoubtedly the greatest cause of failure in flagged footways is the lack of attention paid to the preparation of the subgrade, or to its reinstatement after it has been breached for the repair of services. Differential settlement results in an uneven surface with projecting lips at joints. The condition is aggravated if joints open and allow water to enter, since this results in loss of stability. Movement of slabs will cause the joints to open in this way and allow water to seep through into the base and subgrade, softening them; and should therefore be guarded against. In some instances, because of bad laying, joints are not fully closed and weeds grow through them.
Another cause of failure is that vehicles all too frequently run on pavements adjacent to narrow carriageways and at corners where no precautions are taken; or that cars park with their nearside wheels on the pavement and the side wheels in the roadway. Most authorities, realising the trouble caused by traffic cutting across corners, either lay the slabs on 100 to 150 mm of concrete, or construct the corners with pavings about 150 mm thick. Others now prefer to make up paths alongside narrow roads with bituminous surfaces instead of flags.

6 Kerbs

6.01 Functions

The main functions of kerbs are, **51**:
● To mark the limit of pavement available to vehicles.
● To protect pedestrians and property. For this reason kerbs are usually raised above the surface.
● To provide an abutment against which the drainage of surface water can be controlled.

13 *The dry method of laying slabs: spreading dry mortar base,* **a**; *laying the precast slab and* *checking levels with straightedge,* **b**; *tamping slabs into place,* **c**; *checking levels with straightedge,* **d**.

13a

13b

13c

13d

● To prevent lateral spread of the pavement. In effect the kerb acts as a retaining wall to a granular surface. For this purpose there is no need for the kerb to project above the surface it is retaining—a flush kerb will do the job, **14**; or, as is traditional practice in country districts, the base can be continued under the grass verge for about 300 mm. This will distribute any loads close to the edge and prevent spreading. Concrete slabs carrying light traffic do not, however, need a kerb to prevent spreading and need not be extended to prevent edge deterioration.

6.02 Types

15 shows a range of standard profiles and sizes. Kerbs can be precast to individual designs and colours, but these will necessarily be more expensive. Special finishes can be obtained by exposing the aggregate or facing the kerbs with other materials. Colour can be introduced by the use of coloured cements, the admixture of pigments, and sometimes by the use of special aggregates.

Dowelled kerbs are precast to fit over steel dowels previously fixed in the concrete slab. These kerbs are grouted to the dowels through holes formed in the top surfaces (see paragraph 6.03). True and permanent alignment (withstanding impact) can be obtained by this method, and laying is fast.

Re-kerbing can be economically carried out by using precast concrete facings. These are available with rebated edges which fit over the existing worn kerb (without the necessity for removing the latter). The width of the carriageway is slightly reduced by this treatment.

In addition to the standard products, there is also a range of kerbs for special purposes, **16**. Light reflection, audible warning and illuminated kerbs, for example, are available. Marginal strips, bevelled recesses or projections cast in the vertical faces of the kerbs, are intended to reflect the light of oncoming vehicles at night (and particularly in wet conditions). Reflectors can also be cast into the vertical faces of kerbs. Developments are in hand to incorporate strip-lighting in precast concrete kerbs. This should prove a considerable advantage—particularly at intersections and pedestrian crossings—in bad weather conditions.

Precast 'safety' kerbs are available which present a slightly ramped edge and a 'bulge' profile to the carriageway. These are intended to slow down and bring under control any vehicle which may strike the kerb at an angle at high speed—without 'bouncing' the vehicle back into oncoming traffic. They have a considerable advantage over normal types of impact barrier in other materials, in that maintenance is negligible.

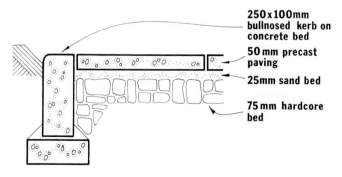

14 *Detail of kerb to paving where it is essential to prevent ingress of water to the base.*

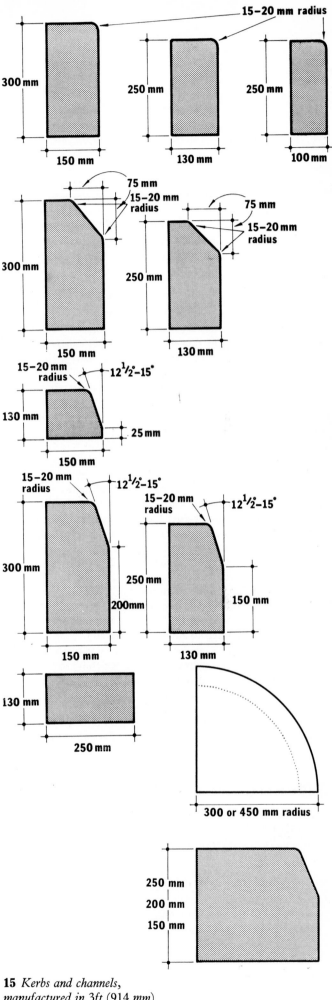

15 *Kerbs and channels, manufactured in 3ft (914 mm) lengths.*

6.03 Laying

It is advantageous to place kerbs *on* the actual concrete slab rather than *alongside* it. This arrangement is not always possible, and recommendations for both methods are therefore given below.

● Kerbs laid on the concrete slab, **17**: the kerbs should be laid flat on a bed of 1:3 cement/sand mortar about 12 mm thick. Increased thicknesses of bedding can be used, especially when there is a change of grade in the surfaces. A recess about 25 mm deep and 300 mm wide is sometimes formed in the slab to take the kerbs, but the constructional difficulties and expense are not usually justified. The kerbs should have a suitable backing to prevent lateral displacement. An effective method involves using 12 mm diameter 150 mm long dowel pins; 75 mm of the length of the pin is fixed into the slab before the concrete has hardened and the remaining 75 mm protrudes into the kerb backing. Two dowel pins should be used for each length of kerb. The line of the dowel pins should be about 50 mm behind the back line of the kerbs. If the kerbs are not to be laid until a considerable time has elapsed after laying the slab, a temporary in-situ kerb 75 mm × 75 mm can be formed over the dowel pins to prevent damage and to control surface drainage before the permanent kerbs are fixed. Later, when the precast concrete kerbs have been laid, extra backing should be provided over and around this temporary kerb. Joints should be made in the kerbs and the haunching in line with the expansion and contraction joints in the slabs. Joints between individual kerbs may be made by leaving an open joint the thickness of a bricklayer's trowel between them or by buttering the whole of the ends of the kerb with mortar (1:3 cement:sand) during the laying to form a completely filled joint.

● Kerbs laid at the sides of the slab, **18**: kerbs should be laid on a concrete base at least 150 mm thick and 300 mm wide and have a concrete backing to within 50 mm below the top of the kerbs. If the kerbs are laid on freshly placed concrete, the provision of mortar bedding (as in the section above) will not be necessary, but, if the concrete has hardened, bedding will be required. The kerbs should abut against the edge of the concrete slab and a longitudinal expansion joint should not be formed in the channel. The kerbs should be jointed to allow for expansion. This may be done by leaving an open joint the thickness of the tip of a bricklayer's trowel between kerbs, or preferably by providing expansion joints. The expansion joints should be placed at 30 m intervals, and should contain a joint filling material such as impregnated fibreboard. The practice of laying kerbs with narrow joints which are pointed afterwards is likely to lead to spalling at the ends of the individual kerbs and is therefore not recommended.

6.04 Recommended details

For non-roadedge situations (eg edgings to pedestrian paved areas, vehicle parking areas, footpaths etc) the details and sizes shown in **14**, **15**, are typical. For road edge details, the following recommendations are taken from the *Guide to good practice for road edge details* (Concrete Society Technical report number 10, published by the Concrete Society, London, July 1974).

In the interests of economy, designs should be as simple as possible. In rural areas, mountable kerbs should be used (that is, splayed kerbs with a total upstand of 75 mm to 100 mm). In urban areas, non-mountable kerbs are usual (that is, vertical or half-batter kerbs with an upstand of about 100 mm and not exceeding 150 mm). The primary function of non-mountable kerbs is to separate vehicular traffic from pedestrian traffic; where such separation is not necessary, it is preferable not to use them as they may

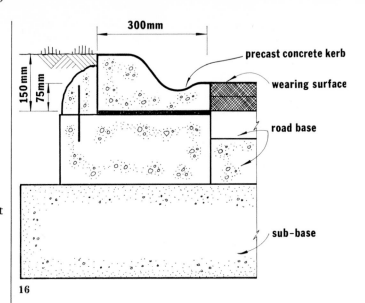

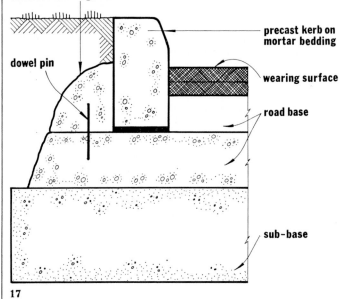

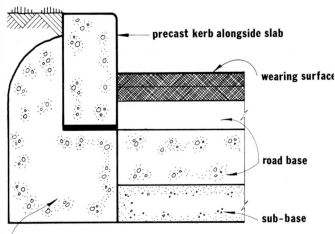

16 *Precast concrete dished 'continental' kerb laid in advance of flexible pavement. Suitable for rural road.*
17 *Kerb laid on concrete slab (recommended)..*
18 *Kerb laid at side of concrete slab (less advisable).*

induce 'kerb-shyness' in motorists (thus decreasing the effective width of the pavement), and may cause skidding vehicles to overturn.

● Urban roads. Unless there is a verge at least 3 m wide between carriageway and footway, a non-mountable kerb should be used. The upstand should be 100 mm in all cases except where a re-surfaceable road is used by heavy traffic, in which case the upstand may be increased to 125 or even 150 mm, to allow for one resurfacing. In most cases half-batter precast concrete kerbs are preferred, **19**. But where there is very heavy industrial traffic, natural stone kerbs might be preferred for their exceptional resistance to abrasion and impact damage; and in situations where there are, in addition, tight radii and where kerbs might frequently be mounted by heavy trucks, steel-faced kerbs might be called for. Extruded kerbs, laid in situ, are not recommended in urban situations owing to the need for fairly frequent adjustment for resurfacing, provision of access and crossing places, and for replacement following damage.

● Housing estate roads. As with urban roads in general, a non-mountable kerb with an upstand of 100 mm (dropping to about 25 mm at vehicular crossings) is preferred. Precast concrete half-batter kerbs are recommended, **19**. Vehicular crossings can be extremely frequent and are facilitated by the use of ramping kerbs, **20**. In many cases it may be felt that natural stone kerbs would be preferable to concrete for amenity reasons; in such cases it should be borne in mind that the cost of stone would be 2·4 to 2·9, if concrete be taken as 1·0 (labour, plant and material included). Vehicle crossings are provided by using half-section vertical kerbs, **22**; and ramping kerbs (left or right-handed) are used to link the crossing kerbs to the main kerbline, **20**, **21**.

● Roads in new towns. Here natural stone or engineering brick kerbs may be preferred to concrete for amenity reasons; and there can be no objection to this, provided the initial capital is available. Where concrete is used, detail **19** would be suitable for the more heavily trafficked areas.

7 Edgings

7.01 Functions

Precast concrete edgings generally fulfil the same functions as kerbs, but are dimensionally much smaller, **23**—to suit the lighter duties for which they are used. They are particularly suitable for defining the edges of footpaths and grassed or paved areas.

Just as kerbs are partially used to keep the road sub-structure dry, so does the edge of the pavings require a seal to keep the water out. This is not absolutely essential when the pavings are laid on stabilised soil or concrete, or where the pavings abut a grass bank or building, or when the edge is at the top of the fall of the paving, but it is usually preferable, **24**. A 50 mm thick precast concrete edging is suitable for this purpose.

The use of concrete edging usually imposes vertical barriers at the sides of the path or paved area. To reduce this effect and increase the depth of base sealed, lay the edging

19 *Full section, half batter precast concrete kerb laid in advance of flexible pavement. Suitable for urban or housing estate road.*
20 *Typical ramping kerb joining a 130 mm × 250 mm half-batter kerb to a 130 mm × 180 mm vertical*

dropped kerb.
21 *Precast concrete kerb laid to leave a 25 mm upstand at a vehicle crossing.*
22 *Half section, half-batter precast concrete kerb laid on rigid pavement. Suitable for urban or housing estate road.*

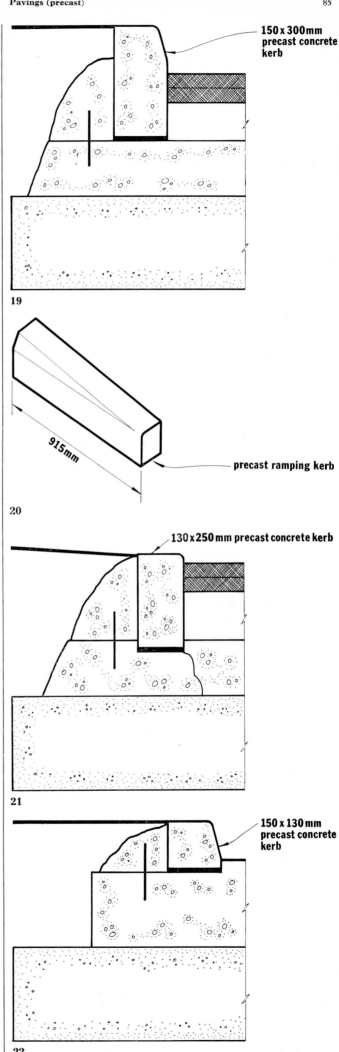

150 x 300mm precast concrete kerb

19

915mm

precast ramping kerb

20

130 x 250 mm precast concrete kerb

21

150 x 130 mm precast concrete kerb

22

as low as is possible in relation to the paving. If water has to be retained within the area of paving, or a flower bed has to be kept off the paving, then a certain height will be required, and if the edging is as low as is possible (say 50 mm) it will be far less apparent, **25**.

8 Channels

8.01 Types and use

In normal use, standard-section precast concrete channels are usually laid against kerbs for the simple purpose of providing the means for rapid and complete drainage of surface water, **26**. Surfacings should be laid slightly proud of the channels to ensure that all water drains off the paving.

In addition to the standard channel sections, there are also special types available for special purposes. For example, channels can be precast with their top surfaces dished or fluted to incorporate gutters, **27**. Channels are also manufactured with hollow cores into which surface water drains through slots formed in the top surface. In some instances the slots have precast metal inserts. Projections on the surfaces of precast concrete channels can be designed to warn drivers that their vehicles are nearing the footpath. The sudden noise made by the tyres on striking these projections gives a timely warning—particularly at night (or in fog)—that the vehicle is tending to run off the carriageway.

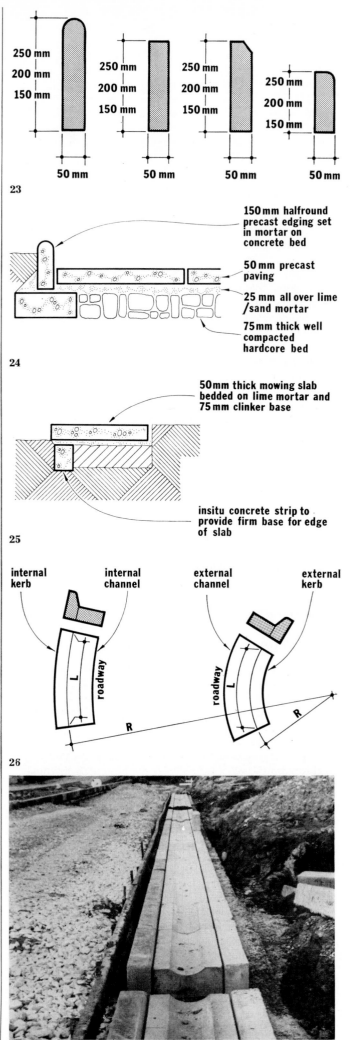

23 *Edgings, manufactured in 3ft (915 mm) lengths. Each of the range of available heights (250 mm down to 150 mm) is shown.*

24 *Edge to planting bed.*
25 *One form of flush edging.*
26 *Radius kerbs and channels.*
27 *Dished channel.*

Information sheet 2
Pavings (in situ)

1 Use of in-situ concrete paving

1.01 Where to use

It will be seen from paragraph 1.03 of Information sheet 1, page 78, that in-situ concrete is a comparatively economical, as well as a durable landscaping finish. Its combination of low initial cost, long life and low maintenance therefore makes it a logical choice for situations in which an economic, impervious, landscaping surface treatment is required.

However, the following precautions should be taken to ensure that the surface will be visually attractive, as well as being functionally satisfactory.

● Do not use in-situ concrete in situations where frequent digging up of underground services will be required (unless removable slabs or manhole covers can be incorporated in the in-situ surface, to obviate the need to break into the paving).

● If large areas are to be covered (as opposed to, for example, a narrow footpath), in-situ concrete can look extremely dull unless broken up by trim, changes of texture, or variation of material. The designer should therefore heed the advice given in paragraph 2.01 of the introductory technical study, and try to organise large expanses of concrete into interesting (and rational) geometrical compositions by means of jointing pattern, introduction of textured finishes, changes of material and level, and so on. Coloured concrete should be used only with great caution.

● Correct detailing, and good workmanship and finishing, is essential. Advice on these matters is given in this information sheet.

In addition to the use of dense concrete for ordinary landscaping, there is the further possibility of *porous no-fines concrete paving* for all-weather play areas. This has been discussed in paragraph 3.02 ('Drainage' page 33) of the introductory technical study and brief advice on installation is given in paragraph 11 of this information sheet.

Finally, there is the use of *cobbles* incorporated as a finish to in-situ paving. This will usually be done in order to discourage pedestrian traffic, but cobbles introduced as a textural variation to relieve the monotony of smooth in-situ paving have great potential. Such areas will usually be relatively small, as they cost about three times as much (in initial installation cost) as plain in-situ concrete, and more than twice as much as a precast paved surface. See paragraph 10.

2 Design considerations

2.01 Bay layout

Bay shape and size
Rectangular bays should, as far as possible, be used in all concrete paving. Although it may appear necessary to use a slab with an acute corner in some instances, this should be avoided as such corners are liable to crack. The layout should be re-arranged to provide only corners with an internal angle of 90° or more.

Where a slab is constructed in two or more bays it is most important that any transverse expansion or contraction joints should be continuous across the width.

Bay width should be of convenient size up to a maximum dimension of 5 m. Each bay can be separated from the adjoining ones by either a plastic sealing or a 5 mm creosoted timber batten although this separating strip is not essential. The limitation of bay size helps to avoid cracking, but if transverse cracks can be accepted, then a bay can be laid as long as convenient.

Arrangement of joints
When deciding the positions of joints the following points should be considered:
● to prevent the development of 'sympathetic' cracking, transverse joints on each side of a longitudinal joint should be in line with one another and not staggered. Transverse joints should always be at right angles to the slab edges; joints at other angles are liable to cause sideways movement of the slabs and diagonal cracking at the more acute corners;
● joints should be arranged so that the use of special forms and very long tampers is avoided as much as possible;
● there should be no acute angles between bays of concrete or corner cracking might occur;
● attention should be given to the position of manhole covers and gulleys, **1**.

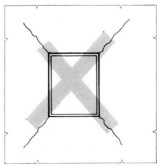

compressible filler board 20mm wide, joint sealed with hot or cold bitumen
longitudinal joint
transverse joint

manhole
(rectangular)

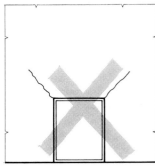

longitudinal joint

manhole
(square)

transverse joint

1 *Incorrect (above) and correct (below) positioning of manhole covers in paving pattern.*

If acute angles in the concrete pavement cannot be avoided, the corners should be strengthened by increasing the amount of reinforcement (see paragraph 5.01).

2.02 Falls

To ensure rapid drainage of rainwater, all slabs must be laid to falls. Transverse drainage is provided by laying the slabs to a straight crossfall of about 1 in 50, with extreme limits of 1 in 30 and 1 in 60. In narrow slabs the fall need only be taken to one side; this reduces the length of surface water drainage required. On wider sections falls to either side of the centre are usual.

The longitudinal fall is usually governed by the topography of the site and by the limitations of the gradients. On land which does not provide a sufficient natural fall, an artificial longitudinal fall may have to be made.

When deciding on the longitudinal fall, the probable standard of workmanship should be kept in mind because small irregularities in the surface may show up as large puddles when the fall is slight. For example, a machine-laid concrete slab built under good supervision could theoretically be laid as flat as 1 in 300 and still give good drainage, whereas a slab laid by semi-manual methods with little supervision should have a longitudinal fall of not less than, say, 1 in 120.

3 Earthworks

3.01 Soil survey

The type of soil on a site has some influence on the design of the slab and base. A soil survey should therefore be made at the start of any construction project to determine the types of soil present and their extent, and also to find the height of the water table.

3.02 Preparation

When the soil has been excavated to formation level any areas of very soft soil or of top soil, which are unsuitable as a subgrade, should be removed and replaced. It may be necessary to compact the subgrade. Granular soils especially may need compaction, as their natural density is sometimes low; if this is because of a high moisture content the soil must be drained before it can be satisfactorily compacted. Undisturbed clays and silts, on the other hand, usually have a very small air content and compaction is unnecessary; in fact, heavy compaction may destroy the internal structure of the soil and result in a loss of strength.

To compact the subgrade, smooth-wheeled rollers weighing from 8 to 10 tons will usually be satisfactory, though with soft soils it may be necessary to use a light roller first.

The subgrade should be protected from the weather both during and after construction, particularly if it is sensitive to moisture variations (eg clays and silts). The prepared formation should never be exposed to the weather as it might dry out or absorb moisture. In general, nor should construction traffic be allowed to run on the subgrade. If, however, the subgrade soil is of good quality (for example gravel, sands, hoggins, stiff clays) and can withstand traffic loads without deformation, traffic can run on it provided it is protected from the weather by an impervious layer. A cheap and satisfactory way of supplying this protection is by surface dressing, ie by applying a bituminous sealing coat to the formation as soon as it has been prepared. An alternative way of avoiding damage to the subgrade is to delay removing the last 100 mm or so of soil above formation level until immediately before laying the base and the concrete slab. This method is particularly suitable where the subgrade is of clay or silty soil.

3.03 Trenches

Wherever possible, trenches should not be constructed in the subgrade. If, however, this cannot be avoided, they should be refilled and fully compacted before the concrete slabs are laid. The filling material should be placed in layers not more than 200 mm thick and each layer should be thoroughly compacted with power rammers or small vibrating rollers. If locally available materials are unsuitable for filling, lean concrete (concrete with an aggregate to cement ratio of about 20:1) can be used.

3.04 Drainage

The purpose of draining the subgrade is to maintain the soil at a reasonably constant moisture content both during and after construction of the concrete slab. Excess moisture weakens a subgrade and change in moisture content means that the foundation might vary in bearing quality.

The subgrade may be affected by the seepage of water from higher ground. Where the surface is constructed on sloping ground, a subsoil drain will usually be needed to intercept water flowing towards the slab. The drains should be back-filled with suitable hard material.

Where there is a possibility of the level of the water table in the subgrade rising considerably above normal, for example in very wet winters, subsoil drains should be installed on either side of the slab.

3.05 Adjacent vegetation

If moisture is drawn from the subgrade, reduction in volume can occur, particularly in clays, which will cause settlement problems. This loss of moisture can occur as a result of the transpiration of moisture by vegetation growing close to the slab.

4 Bases

4.01 Purpose

A base is included to provide a hard smooth working face over the formation and it is usually prepared to an accurate surface level to ensure a uniform thickness of slab. In addition, the base will protect the subgrade during construction of the concrete slab.

4.02 Base formation

Hoggin, well-graded gravels or sands, lean concrete, cement-stabilised soil, or artificial materials such as hard clinker or slag can be used as base materials. Hardcore may also be used, but it should be crushed so that it is graded from 50 mm down, see table 1.

Large lumps of hard material such as brick or concrete are unsatisfactory; ashes are also unsuitable as they are unstable when wet.

The top surface of the base should be finished and rolled to a smooth level surface. Fine material should be added, if necessary, to close the voids; sand will be satisfactory for this purpose, and has the additional advantage of reducing friction under the concrete slab to a minimum, thus allowing it to move freely with changes in temperature. It may sometimes be necessary to allow traffic on the base before the concrete slab is laid. In this case the base should be surface dressed in the same way as the formation to prevent the admission of water and to avoid surface damage.

Table I Types of subgrade and recommended base thickness

Subgrade	Materials	Base thickness
Normal	Well graded and drained: ● sand ● gravel ● hoggin ● stiff clay	75 mm
Weak	Clay Silt Sandy silty clay with high water table	150 mm

4.03 Slab thickness

Slab design is very much dependent upon two factors— first the quality of the subgrade and secondly the nature and amount of traffic expected to pass over the surface. Not only will most hard landscape need to be designed for pedestrian use but increasingly it will be required for service vehicle access, temporary storage facilities and the transfer of goods by various mechanical means. Table I indicates the types of materials usually encountered in subgrades and these are classified as either normal or weak. Table II indicates likely ranges for both static and live loads and depending upon the types of subgrades encountered provides overall guidance on the likely slab thickness which would be suitable for most circumstances. Further guidance on concrete quality is found in Table III.

Table II Guidelines for slab thickness for various conditions of static and live loads

Typical loading (static)	Subgrade	Slab thickness
Light (up to 5kN/m²)	Normal Weak	150 mm 175 mm
Heavy (between 5 and 20kN/m²)	Normal Weak	175 mm 200 mm
Typical loading (live)		
Light (up to 10 000 kg)	Normal Weak	175 mm 200 mm
Heavy (between 10 000 and 30 000 kg)	Normal Weak	200 mm 250 mm

5 Reinforcement design

5.01 Factors affecting design

Although the quality, or bearing capacity, of the subgrade soil makes little difference to the stresses induced in the concrete slab, the intensity and type of traffic using a concrete slab do affect its performance considerably. It is these factors that are therefore used as the basis for the design of the slab. See table II.

When concrete slabs are unreinforced, fine cracks that appear in them sometimes widen progressively until finally they may open sufficiently for interlock between the aggregate particles on either side of the crack to be lost. Foreign matter and water can infiltrate into the cracks and cause weakening of the subgrade.

When concrete slabs are reinforced, 2, however, the reinforcement holds together the edges of any cracks that may appear, so that they remain harmless hair cracks. In addition, the use of reinforcement permits wider spacing of joints. Reinforcement is not intended primarily to prevent cracking or to increase the load-carrying capacity of the slabs, though it fulfils these functions to a limited extent.

Acute angles

If acute angles in the concrete pavement cannot be avoided, the corners should be strengthened by increasing the amount of reinforcement. This may be done by using two 16 mm diameter bars, each bent in the form of a hairpin with arms 1-5 m long and with hooked ends. The hairpins should be placed at the mid-depth of the slab with the arms arranged radially at approximately equi-angular intervals. Alternatively, an additional mat of square mesh reinforcing fabric, not less than 1 m square, should be placed in the corner, at the top of the slab.

5.02 Placing of reinforcement

Experience has shown that the number of layers and the position of the reinforcement in the slab are not critically important. From the practical point of view, however, one layer is preferable because double reinforcement is more difficult to lay properly and entails a greater number of operations; it is therefore more expensive, weight for weight, than single reinforcement.

The single layer of reinforcement should be placed about 50 mm from the top of the slab, 4. This position is recommended for the following reasons.

● Steel near the top will provide a greater insurance against the opening of cracks in the surface than will bottom reinforcement.

● Full compaction of the concrete is easier to obtain if the steel is placed near the top of the slab.

● Reinforcement placed in the lower position is open to attack from moisture in the subgrade if, through bad workmanship, the concrete is honeycombed at the bottom of the slab.

Generally the longitudinal bars will be placed parallel to the main axis, 3. However, in some cases, for example where alternate bay construction is used, the width of the slab will be greater than the length when measured along the axis; the reinforcement should then be placed with the longitudinal bars at right angles to the main axis. When placing the single layer of reinforcement, the following procedure is recommended, 4. A layer of concrete is spread on the formation and struck off and compacted to the level designed for the reinforcement. It is laid for a length slightly exceeding that of the reinforcing mats, which are then placed in position overlapping at least 300 mm at the transverse joint; lapping at longitudinal joints is not necessary. Wire ties should be used to connect the mats as these help the steel to lie flat. The surface layer of concrete, which may be the same as that placed in the lower layer or may be a special mix, should then be spread, as soon as possible after the spreading of the lower layer as the placing of the reinforcement will allow; this ensures that the bond between the two layers is complete. In no circumstance should more than 30 minutes elapse between the placing of the two layers; for this reason, it may sometimes be necessary to restrict the length of the reinforcing mats used.

2a

2b

3

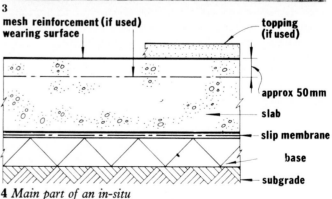

4 Main part of an in-situ slab.

2a Fixing reinforcement in position 50 mm below final level of paving; b ready-mixed concrete being fed on to reinforcement by chute, for spreading, levelling, and finishing by mechanised paver in background.

3 Longitudinal bars being fixed parallel to side forms.

If, for any reason, it has been difficult to backfill trenches in the subgrade satisfactorily, extra reinforcement should be provided in the slab over the affected area. It should be placed 50 mm above the bottom of the slab and care should be taken to ensure that the concrete is thoroughly compacted round it. The fabric should extend about 450 mm beyond the edges of the trench and should be placed so that the bars containing the majority of the steel weight lie across the trench.

6 Concrete design

6.01 Strength

The higher the strength of the concrete in the slab, the better will be the performance, but the question of cost limits the strength which can be used in practice. It has been found that concrete having a minimum cube crushing strength of 21 N/mm^2 at 28 days is desirable. On a very small job, the strength may be neither specified nor measured but the concrete should have this minimum strength to ensure satisfactory durability. Experience has shown that if a concrete of this strength is used, other essential requirements, such as resistance to wear, will also be obtained. See table III.

Table III Guidelines for concrete quality	
Duty	Grade N/mm^2
Pedestrian access only	21
Limited service access and light storage	25 to 30
Heavy duty vehicle access and heavy storage	40 or higher

The average strength of the concrete needs, of course, to be higher than the minimum figure given above because of the need to allow for variations in materials and mix proportions.

6.02 Air entrainment

Well-made, dense concrete is highly resistant to the effects of frost; where, however, neat salt is used to disperse snow, ice and frost, the disruptive effect of frost is considerably increased. For this reason it is recommended that an air entraining agent should be used whenever it is likely that neat salt will be subsequently applied to the concrete surface. The total air content required is $4\frac{1}{2}$ per cent by volume, with a margin of $\pm 1\frac{1}{2}$ per cent to allow for site variations. The quantity of air entraining agent used should be as advised by the manufacturer (see Information sheet 7).

6.03 Workability

The workability of the concrete must be high enough to ensure that a fully compacted concrete is obtained, together with a good surface finish. If the concrete is too dry and compaction inadequate, the slab will be honeycombed and will not be durable. If, however, the concrete is too wet—as often occurs on construction sites—the slab will have low strength and the surface will be liable to become smooth and to wear rapidly. The workability, moreover, must not be so high as to cause the concrete to flow down the crossfall or camber during compaction. If the crossfall is normal, ie 1 in 50 to 1 in 60, there should be no difficulty, but where there is higher elevation or on steep gradients care must be taken to ensure that the degree of workability is

only just sufficient to produce full compaction. The final choice of degree of workability is usually assessed on site; so is the decision on whether admixtures are appropriate.

7 Joint design

7.01 Functions and types

As concrete expands and contracts with changes in temperature and moisture content, concrete slabs can warp if the temperature or moisture content varies through the depth. It is to prevent excessive stresses resulting from these causes that joints are made in concrete slabs—their function being to permit movement to take place under controlled conditions.

Expansion joints, contraction joints, longitudinal joints and construction joints are those most likely to be used. Expansion joints are provided to allow the concrete to expand beyond its original length. Contraction joints prevent or control cracking caused by tensile stresses in the slabs when expansion joints are widely spaced. In most circumstances, longitudinal joints will be necessary where the slab is more than 5 m wide. Construction joints should only be made when unavoidable interruptions occur in the progress of the work.

Expansion joints

The cavity of an expansion joint, **5**, which extends to the full depth of the slab, is filled with a material which will easily compress and is sealed at the top with a plastic compound to prevent the ingress of water. The filling and sealing materials together should extend to the full depth and width of the slab; the joint should be vertical and form a complete break between adjoining slabs.

Dowel bars included in an expansion joint enable the slabs to give support to one another and prevent small irregularities appearing at the joints because of differential settlement. In addition they provide a means of transferring heavy traffic loads from one slab to another. Dowel bars should not be provided in slabs less than 150 mm thick, as the amount of concrete cover would not be sufficient to prevent spalling over the bars.

Contraction joints

The layout of a contraction joint is shown in **6**. The joint consists of a groove cut or formed in the slab to weaken it, so that if a tendency to crack occurs, the position of the crack is predetermined. Such cracks are then easily kept sealed and do not detract from the appearance of the surface.

Construction joints

Construction joints, **7**, are made when unavoidable interruptions occur in the progress of the work, because of plant breakdowns, adverse weather conditions or other unforeseen causes. The day's work should normally begin and end at an expansion or contraction joint, and construction joints should be regarded as emergency joints and avoided whenever possible.

Construction joints should not be placed nearer than 3 m to an expansion or contraction joint. If construction joints must be made, they should take the form of a butt joint, with the two portions tied together by carrying the reinforcement through. On poor subgrades it is desirable also to provide tie bars across the joint. These should be 10 mm diameter straight rods 1 m long, and placed at 450 mm centres. A sealing groove should be formed at the top of the joint.

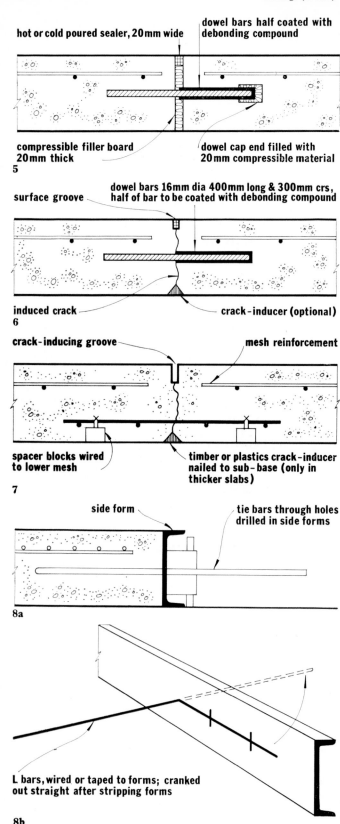

hot or cold poured sealer, 20mm wide

dowel bars half coated with debonding compound

compressible filler board 20mm thick

dowel cap end filled with 20mm compressible material

5

surface groove

dowel bars 16mm dia 400mm long & 300mm crs, half of bar to be coated with debonding compound

induced crack

crack-inducer (optional)

6

crack-inducing groove

mesh reinforcement

spacer blocks wired to lower mesh

timber or plastics crack-inducer nailed to sub-base (only in thicker slabs)

7

side form

tie bars through holes drilled in side forms

8a

L bars, wired or taped to forms; cranked out straight after stripping forms

8b

5 *Expansion joint;*
6 *contraction joint. Structural consultant should advise on joint spacing. But as very rough guide, reinforced slab for pedestrians plus occasional light vehicles may have contraction or construction joints at 15 m, and expansion joints at 45 m.*
7 *Construction joint, provided at points of interruption to the concrete-laying process.*

8 *Longitudinal joint, with 1 m long, 10 mm dia tie bars at 750 mm centres, to hold slabs together. Two alternative methods of casting in the bars are shown in 8a and b. Top of joint should have a sealing groove, filled with sealing compound.*
9 *Sawn contraction joint.*
10 *Impregnated fibreboard joint filling being placed in position while concrete is in plastic state.*

Longitudinal joints

Longitudinal joints, 8, prevent the formation of longitudinal cracks which experience has shown tend to occur in slabs more than 5 m wide. Also if slabs are more than 5 m wide, with manual techniques the tamping beam becomes heavy and difficult to handle. It is recommended that tie bars should be provided in the longitudinal joints at mid-depth of the slab to reduce the possibility of differential vertical movement between the slabs and to prevent the slabs from creeping apart.

Sawn joints

A comparatively recent introduction has been the technique of sawing joints after the concrete has hardened, 9. Water-cooled circular blades of various types are used. The engine, driving mechanism and blade are mounted on wheels in a convenient portable frame which can be propelled along the road surface.
The saw can be used for forming a sealing groove at a longitudinal joint, for forming a contraction joint and also for sawing a groove to contain the joint sealing compound over the preformed jointing material in an expansion joint. The cost of sawing joints is marginally higher than hand methods but it saves labour, gives a better finish, and the evenness of the surface is not so dependent on manual skill.

Joint fillers and sealers

Suitable materials for filling joints are soft knot-free timber, impregnated fibreboard, 10, chipboard, cork and cellular rubber. Materials containing a high proportion of bitumen should not be used: they are likely to extrude.
A joint sealing compound should be used to protect the joint against the entry of water and grit. The most frequently used sealing compounds are rubber-bitumen mixtures, oxidised bitumen compounds, straight-run bitumen compounds containing fillers and resinous compounds.

9

10

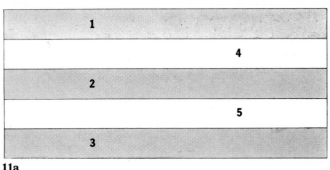

11a

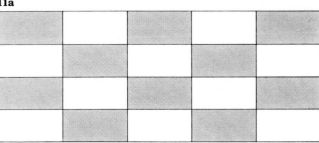

11b

8 Construction

8.01 Continuous and alternate-bay methods

In the continuous method of concrete construction, **11a**, concreting proceeds throughout the day and any joint-separating materials and load-transfer devices are incorporated in the concrete as the work proceeds. Alternate bay construction, **11b** involves leaving intermediate bays to be filled in after the concrete of the first bay has hardened sufficiently. Continuous construction is preferable to alternate bay work because it is usually quicker, more efficient, provides better access, and the results are better.

8.02 Underlays

Waterproof paper or a plastic film should be used under a concrete slab, except where there is a hard smooth surface to the base. This prevents the escape of cement paste into the base, reduces the friction between the concrete slab and the base and prevents any soil or other loose material from becoming mixed with the lower layers of freshly placed concrete.

The underlay should be laid flat, free from crinkles and holes, and with adequate laps (about 100 mm) at the edges and ends of sheets. Care must be taken to ensure that the underlay is not damaged after laying and before covering with concrete.

8.03 Formwork

Steel forms are now almost invariably used and are available in a wide variety of sizes and types. Those with a wide flat base are preferable since they are more stable and less likely to move under the action of the vibrating beam. Timber forms are not generally recommended.
The forms should be fixed to the base, **2a**, using a sufficient number of steel pins of adequate length to ensure rigidity, and the interlocking devices should be driven fully home. Forms should not be supported on packing at the ends but should be laid on a continuous bed of cement mortar or lean concrete about 50 mm thick; they should be laid slightly high and brought to level by tapping down. As the concrete surface is finished by a tamper which is levelled from the top of the forms, it is essential that the forms should be correctly aligned by boning over lengths not exceeding 15 m, or by using a 15 m long stretched line. Forms should be laid at least one day ahead of concreting. Every form should be checked before use, and any form on which the top profile departs from a straight line by more than 1 mm should be rejected. Care should be taken in cleaning and handling forms so that they are not damaged.

8.04 Transporting and placing

Whatever method is used for transporting the concrete, it is important to ensure that segregation does not occur. Segregation is less during vertical tipping than when the concrete is allowed to discharge down a sloping chute. It rarely occurs during transport but it can be prevented by adjusting the mix, by reducing the length of haul or by using a more softly sprung vehicle, or by a combination of these methods. During very dry or very wet weather, it may be necessary to cover the concrete with tarpaulins while in transit.
The concrete will usually be placed by hand. It is common practice to dump the concrete from the transporting vehicle on to the formation and to level it in readiness for compaction. This procedure frequently causes difficulty in

12

11a *Continuous method of laying concrete (numbers indicate sequence); and* **b** *alternate-bay method.*
12 *Mechanised concrete road construction; note steel side forms.*

producing a good surface; the concrete in the bottom of the heaps becomes compacted during the tipping operation and is then much denser than adjacent concrete which is shovelled into position from the upper part of the heaps, and when vibrated, 'high spots' tend to occur over the pre-compacted areas. The concrete should therefore be tipped on to a banker board either to the side or on the formation, and the material shovelled into position by hand; the initial compaction of the concrete is thus uniform.
To allow for the subsidence which occurs during compaction, the concrete must be spread to a surcharge. The amount of this extra thickness will depend on the concrete mix, but for normal concrete it will generally be about one-fifth of the depth of the concrete bay when compacted. A light screed working on timber battens clipped to the top of the forms simplifies the striking off to a uniform surcharge height.
It is more difficult to achieve an even surface when tamping off kerbs than when tamping off forms. Whenever a surface with a high standard of surface finish is required, tamping should be off forms.

8.05 Compaction

A number of devices are commercially available for compacting concrete by the semi-manual method:

● pneumatic compactors: these will satisfactorily compact concrete of low workability but they do not produce a high quality surface finish and a special pneumatic finisher is normally also needed;

● vibrators fixed to a wooden beam, preferably shod with a steel channel, as this gives the best results. Where mains electricity is not available, small petrol- or diesel-driven generators can be obtained to operate these machines;

● vibrators mounted on steel beams, **13**. For convenience the length of the beam should not exceed 4 m; 5 m is the maximum length that can be handled with any degree of ease. Unless the units are mechanically or electrically interlocked so that they vibrate in phase, it may be harmful to use more than one vibrator in one beam.

Whatever type of vibrator is chosen, great care must be taken to ensure complete compaction. The beam is lifted on to the surface of the uncompacted concrete which has been laid to a surcharge above the forms. It is then allowed to sink down to form level under its own weight and the effect of the vibrating unit. After the strip of concrete beneath the beam has been compacted, it should be lifted up and moved forward about 50 mm, assuming a 75 mm wide beam. The next strip of uncompacted concrete is then dealt with and the process repeated, **14**. Thus a fully compacted slab with a slightly ridged surface will be obtained. The concrete should never be spread level with the tops of the forms or allowed to move along under the propelling action of the vibrators. This may produce a satisfactory surface but the concrete will not be densely compacted and the slab will therefore be weak.

In reinforced concrete slabs the concrete has to be spread in two layers to enable the reinforcement to be placed accurately. It is not always essential to compact the two layers of concrete separately. However, two-layer compaction is usually necessary in slabs 200 mm or more thick, and is recommended for thinner slabs also as it produces better quality. One or two passes with a hand vibrator will generally produce satisfactory compaction of the lower layer.

Finishing should be carried out continuously over as long a stretch of concrete as possible. If pneumatic compactors are used, a separate machine is necessary to finish the surface. These finishers have a smaller amplitude and operate at a higher frequency than the compactors and produce a smooth surface.

Mechanically driven vibrators can be used for finishing the concrete as well as for compaction. The finishing pass should be made close behind the compacting operation and the beam should be tilted slightly so that the leading edge is about 10 mm higher than the trailing edge, to facilitate the striking off of any small quantities of excess concrete. When vibrators are used to obtain the final surface finish they should not be allowed to remain too long in one position, and too many passes of the vibrator should not be made, as this will bring an excessive amount of fine material and laitance to the surface of the concrete.

When relatively large outputs of finished surface are required, consideration should be given to the use of a small mechanical finisher. This machine strikes off the surface of the spread concrete to give the correct surcharge, compacts the concrete with a vibrating beam and finishes the surface with a reciprocating screed.

It is often difficult to remove all irregularities in a concrete surface with a hand-held vibrating beam, but the use of a scraping straightedge will ensure a surface well within the usual specified tolerance of 3 mm in 3 m.

It is difficult to use the scraping straightedge satisfactorily if the concrete mix is too harsh or dry. Immediately after the concrete has been compacted the straightedge should be drawn across the surface. The number of passes made should be kept to the minimum necessary to give a smooth surface. The straightedge is moved along the concrete slab by a distance about 1 m less than the length of the straightedge and the operation repeated. Any tearing of the surface can be remedied by another pass of the vibrating beam.

The time allowed to elapse between mixing the concrete and final finishing in place should not normally exceed one and a half hours. A most satisfactory final surface finish is obtained by brushing the concrete with a medium-soft broom after the final pass of the vibrator or scraping straightedge. The brush should be drawn across the surface in one continuous sweep.

Special surface finishes can be obtained where a decorative finish is required; methods of producing these are given in Information sheets 9 to 16. On very steep slopes, a slightly ridged surface, as produced by the vibrating beam, is advantageous to improve traction in frosty weather.

8.06 Curing

Concrete which is allowed to dry out quickly loses strength and wearing qualities, and surface cracks may develop. Newly completed concrete slabs should be cured and protected from sun and wind so that the water in the concrete does not evaporate. This protection is also necessary to prevent the surface from being pitted by heavy rain which might fall before the concrete has hardened.

Curing should commence as soon as possible after the concrete has been laid; no slab should be left unprotected for more than 15 minutes after finishing. Curing should continue for seven days in warm weather, but in cold

13 *Tamping concrete by means of vibrator fixed to steel beam.*

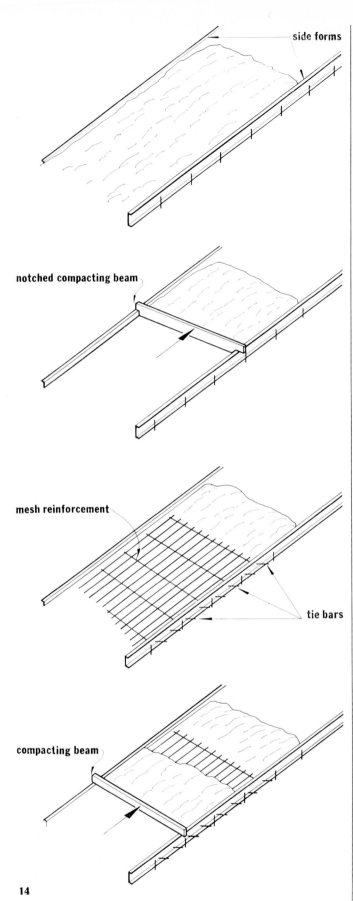

side forms

notched compacting beam

mesh reinforcement

tie bars

compacting beam

14

15a

15b

15c

14 *Stages of concrete laying. First layer of concrete is laid between steel side forms; concrete is compacted by tamping beam; reinforcement is fixed 50 mm below finished level; top layer is laid and* *compacted/smoothed. See also* **2** *and* **4**.
15 *Curing methods, showing spraying,* **a**; *tarpaulins on frames,* **b**, **c**.

weather this period should be extended. In frosty weather the number of days of frost should be added to the curing period given above. The following materials can be used for curing insitu concrete surfaces.

● damp fabric, such as hessian—when placing the fabric, care should be taken to avoid marking the surface of the concrete. The fabric should be kept damp throughout the curing period, **15abc**;

● damp sand—this cannot normally be applied until the day after the concrete has been laid and must be kept damp throughout the curing period. Before this, some other curing process must be used, such as a waterproof covering on a light framework;

● damp straw or reeds—this is applied in the same way as damp sand and must be kept damp throughout the curing period. It is advisable to make the material into

mats; this can be done using a suitable thatch-making machine, or purpose-made mats can be purchased.
● waterproof coverings such as paper or tarpaulins—these make excellent curing materials provided they form a close cover that completely excludes wind. Light coverings such as paper can be applied direct to the concrete very soon after laying but they must be well fastened down at the edges. Waterproof paper made into full width curing blankets, which are easy to handle and which can be re-used many times, can be obtained commercially. Heavier coverings, **15b, c** such as tarpaulins may have to be supported above the slab to avoid marking the surface, but they should extend over the edges of the slab to exclude draughts.
● plastic film—Polythene or polyvinyl chloride film is widely used for curing purposes. It is very light and therefore easy to handle. It can be applied direct to the concrete, but to prevent the formation of a smooth glazed finish, care must be taken not to place it in contact with the wet surface.
● curing membranes **15a**—a number of proprietary curing membranes are available which can be sprayed on to the surface of the concrete as soon as free water brought to the surface by the finishing process has evaporated. It is important that the rate of spread be uniform and the coverage in accordance with the maker's instructions. It is preferable to make two applications to ensure complete coverage of the surface. Bituminous membranes are not recommended as they may produce a slippery surface and they take some time to wear off, leaving a blotchy appearance. For coloured concrete work the curing should be done with waterproof paper or plastic sheeting, as other materials are liable to stain the surface.

9 Winter construction

9.01 Precautions

Unless precautions are taken it is inadvisable to lay concrete when the temperature is below 4°C. Concrete slabs can, however, be constructed satisfactorily in winter if precautions are taken to prevent damage to the formation and to the concrete arising from excessive moisture and/or low temperatures. It may not be economical to adopt special measures to enable concreting to continue in bad weather, but if it is necessary to continue concreting, the precautions set out below should be observed. Precautions 1 and 7 are particularly important.
1 Frozen aggregates must not be used.
2 The aggregates may be heated, for example by passing steam through the stockpiles.
3 The mixing water may be heated to a temperature of up to about 82°C.
4 Extra-rapid-hardening Portland cement may be used, or if rapid-hardening or ordinary Portland cements are used calcium chloride may be added: the amount of commercial grade calcium chloride used must not exceed 2 per cent by weight of cement.
5 Richer mixes may be used.
6 The subgrade should be protected so that it is not frozen when the concrete is laid on it.
7 Straw or other insulating material should be placed over the concrete as soon as possible after finishing. It is convenient to cover the slabs first with a layer of waterproof paper, followed with a layer of dry straw 150 mm to 300 mm thick, either loose or in blankets. A final covering of a second layer of waterproof paper should be given. In conditions of light night frost, precautions 1, 4 and 7 only are usually sufficient. When the temperature

does not rise above freezing point by day or night it will be necessary to observe precautions 1, 3, 6 and 7. In conditions of prolonged and severe frost it may be necessary to heat the subgrade with braziers, to heat the mixing water and perhaps also the aggregates so that the temperature of the concrete at the time of placing is not below 15°C, and to use a layer of insulation over the concrete. To enable the chemical action of hardening to proceed it is essential to keep the temperature of the concrete when in the plastic state at least 3°C above freezing point.

10 Cobbles

10.01 Where to use

In addition to the attractive-looking appearance of cobbled paving materials and one whose small scale and deep texture have not been imitated by any new material, there are many practical reasons for its use.
By the judicious use of cobbles, pedestrians can be discouraged from cutting corners, straying on to the road or over grass. Cobbles are also very handy for small areas and 'left over' spaces which have awkward shapes in plan or section. They are small enough to fit in where a paving slab might be difficult to cut, or where a plain surface is impossible. Similarly, loose cobbles can be useful around tree trunks, particularly if for some reason the level of the ground has to be raised. They will allow air as well as moisture to penetrate to the roots.

10.02 Patterns

Cobbles can be set at random, roughly coursed, or laid in patterns such as the traditional Welsh herring-bone. This depends to some extent on the shape of the actual stone; the long flattish stones when laid on end or edge have a sense of direction and are more suitable for pattern-making. The most satisfactory looking cobbled paving is that where the stones touch and the gaps between them are as small as is possible so that the concrete in which they are set is out of sight. Far less satisfying is that where the concrete is nearly flush with the stones so that it is as visually important as the cobbles. Another traditional method rarely seen now is that where a pattern of small cobbles is laid in a mass concrete bed. This might almost be classed as an exposed aggregate, if it were not for the deliberate placing of the stones.

10.03 Methods of laying

For pedestrian areas on a well consolidated foundation, 50 mm of compacted sand is spread. On this is laid about 50 mm of concrete (1:2:4 mix with a small aggregate). The cobbles are then pressed in by hand until they protrude the amount required, **16**. The final thickness of the concrete will be about 75 mm. A traditional method is to lay the cobbles in sand or fine gravel on a hard foundation and ram them in.
The Italian method is to bed the cobbles in dry concrete or mortar, with the cobbles touching and protruding so that the bed cannot be seen. In order to avoid marking the stones, water is added after the pavement is laid.

16a

16b

16c

11 No-fines self-draining paving

11.01 Introduction

The demand for all-weather play areas is steadily increasing, and concrete has many advantages to offer. Properly constructed, it has a long maintenance-free life. A self-draining material with a long life is a very attractive proposition in a country where rain falls frequently throughout the year. A paving constructed of non-fines concrete, **17**, meets this requirement and has a great deal to recommend it, provided care is taken to prevent gradual blocking (silting up) of the voids.

11.02 Site preparation

The existing ground should be excavated to the required levels and the formation prepared so that water draining through the no-fines concrete can drain away as quickly as possible. When using base types 1 and 3, a 500-gauge polythene sheet should be laid over the base to form an impervious layer and the surface finished with a fall of not less than 1 in 80 to allow the water to drain off to land drains around the perimeter of the area. If the area is large, intermediate drainage lines may be needed.
In the case of base type 2 (no-fines concrete), a polythene sheet is placed below the base and the formation graded to a slope of 1 in 80 and discharges the water also to perimeter drains.

11.03 Bases

Base type 1
100 mm of compacted hardcore or granular material such as hoggin.

Base type 2
100 mm of no-fines concrete with an aggregate/cement ratio of 6. The water/cement ratio should be approximately $0 \cdot 42$. It is suggested that the aggregate for this sub-base course should should be 60 per cent 40 mm single size, and 40 per cent 20 mm single size.

Base type 3
100 mm of lean concrete with an aggregate/cement ratio of 18. The water/cement ratio should be between $0 \cdot 42$ and $0 \cdot 50$. This base should be well compacted by a heavy vibrating hand guided roller. It is usual to give one to two passes with the roller without vibration and then the vibrator is switched on until it is seen that the concrete is thoroughly compacted. Maximum size aggregate of 40 mm should be used. **18** shows sections through play areas with the three different sub-bases.

11.04 Methods of laying

The top no-fines concrete slab should be laid in two courses, monolithically. The lower course is approximately 60 mm compacted thickness, using 10 mm single size aggregate, and the top course 40 mm compacted thickness,

16 *Cobbles being placed in mortar by hand,* **a**; *tamped down by timber beam working off side-forms,* **b**; *brushed off,* **c**.

using 5 mm-2·40 mm ($\frac{3}{16}$in-no 7) single size aggregate.
The latter aggregate is non-standard size, and may require
to be specially sieved out. These two courses should be
compacted separately, but the top course must be laid
within two hours of the laying of the lower course.
The mix for both courses should have an aggregate/cement
ratio of 4, and it will probably be found that the
water/cement ratio will be rather higher than in the base
course and may vary from 0·45 to 0·55. The courses
should be compacted by means of a rolling pipe. The pipe
can be 300 mm diameter and either steel or cast iron. It
may be necessary to compact the edges of the slab against
the forms by means of a heavy punner. It is important to
keep the pipe wet. Care must be exercised to avoid closure
of the open porous texture of the no-fines concrete. In
particular water content and compaction are critical features.
Ordinary Portland cement should be used, but a coloured
Portland cement can be used for the top course of the top
slab. The use of admixtures is not recommended. The
aggregate can be gravel or crushed rock, whichever is the
most readily available. Reinforcement is neither required
nor should it be used.
Bay size should be of convenient size up to a maximum
dimension of 5 m. Each bay can be separated from the
adjoining ones by either a plastic sealing strip or a 5 mm
creosoted timber batten, although this separating strip is not
essential. The limitation of bay size helps to avoid cracking,
but if transverse cracks can be accepted, then a bay can be

laid as long as convenient. Both the sub-base concrete (if it
is used) and the top should be thoroughly cured for seven
days by being kept covered with a polythene sheet. It is not
advisable to do any concreting work in very cold or frosty
weather.

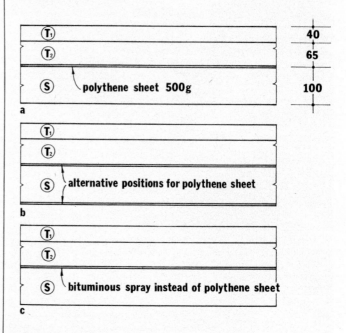

18 *Play areas*
T_1 *Top course of 5 mm single size aggregate.*
T_2 *Top course of 10 mm single size aggregate.*
S Base: type 1, **a**; *type 2,* **b**; *type 3,* **c**.

17a

17b

17a *Porous surface texture of*   *formed to coincide with white*
no-fines concrete. Joint   *lines of tennis court,* **b.**

Information sheet 3
Banks, slopes
and retaining walls

1 Soil tests

1.01 Procedures

It is a prerequisite of building, to ascertain the nature of the ground by digging trial holes or test pits. The number of these trial holes will depend on the nature of the site and the soil variation encountered: two or three should be regarded as a minimum, and they are best located just outside the perimeter of the area to be landscaped, to prevent unnecessary backfilling.

Exploration can be carried out using an earth auger (a boring tool like a large corkscrew); or by driving in a tube of about 40 mm diameter and extracting a core of soil for examination; or by partial pit digging followed by auger or tube tests.

Simple loading tests can be carried out on the soil thus extracted by applying a suitable force per unit area and measuring the settlement with a level. In cases where the magnitude of the problem seems to warrant a thorough survey and soil analysis (quite usual in the case of foundations for buildings, but less likely in the case of simple landscaping) the *AJ Handbook of building structure*, Section 4 (pages 99-156)* should be consulted, or the advice of a structural engineer obtained.

1.02 Soil types

A primary distinction is made between *topsoil* (a surface layer of soil consisting mostly of decayed vegetable matter able to support plant life), and *subgrade* (all soil strata lying beneath the top layer of organic soil).

The subgrade can in turn be classified into one of two categories:
● cohesive (plastic) soils—eg fine-grained silts and clays whose shear strength is derived from resistance due to tendency of particles to stick together;
● non-cohesive (non-plastic) soils—eg coarse-grained or granular soils, including gravels and sands, whose shear strength is derived from resistance due to friction between the particles.

The general nature of a subgrade on a particular site will normally be known from local construction experience and on small, simple contracts this information may be sufficient, with test borings and soil analysis needed only where it is unavailable or where there are special problems.

Subgrade behaviour
Where paving is founded on clays (ie *cohesive* soils) or soil with an underlying layer of clay, settlement is a long-term

* Published as a book by The Architectural Press (London), £4·50 (paperback).

process, and will depend on the nature of the clay and its depth and moisture content—the greater the load, the greater the depth of clay, and the greater the moisture content, the greater will be the settlement.

Most *non-cohesive* subgrades pose few problems, being generally stable and having a fairly high bearing capacity. They are easier to compact than cohesive soils, are unlikely to shrink or swell, and are not susceptible to frost heave. Moreover any settlement that does take place with non-cohesive soils will occur almost immediately after the full load has been applied.

In special circumstances it may be desirable to alter the properties of the subgrade, as an alternative or in addition to other methods of foundation design. The processes available are part of the science of geotechnics, and include making use of cement, clay, silicates and bituminous emulsions; ground-water lowering by deep and shallow wells and well points; base exchange; deep and shallow compaction; soil stabilisation; and the use of compressed air. Specialist advice is essential.

2 Bank and slope design
2.01 Precautions

Provided the subgrade is fully stabilised and paving is carried out within the general limits set out in Table I, then the procedures set forth in Information sheets 1 and 2 for precast and in-situ pavings should be followed.

In all instances it is crucial to take steps to ensure that surface water does not penetrate to subgrade level—eg by providing intercepting drains at the tops of all banks.

In addition to the danger from water, there is also the presence of sulphates in the soil to guard against. The presence of sulphates in either topsoil or subgrade can cause deterioration in concrete unless the precautions given in Table II are taken.

Table I Working classification of soils

Type	Condition	Characteristics	Approximate angle of repose	Bearing capacity* kN/m²
I Rock	Solid	Requires at least a pneumatic or other mechanically operated pick	50° plus	650 plus
II Gravel Sand Rock waste	Compact	Requires pick for excavation. 50 mm wooden peg hard to drive in more than 100 mm	40°-50°	430-650
III Clay Sandy clay	Stiff	Cannot mould with fingers. Requires pick, pneumatic or other mechanically operated spade for removal	35°	220-430
IV Clay Sandy clay	Firm	Can mould with substantial finger pressure and can be excavated with graft or spade	25°	110-220
V Sand Silty sand Clayey sand	Loose	Can be excavated with spade. 50 mm wooden peg can easily be driven in.	20°	110-220
VI Silt Sandy clay Silty clay	Soft	Fairly easily moulded in the fingers and easily excavated.	15°	55-110
VII Silt Clay Sandy clay Silty clay	Very soft	Natural sample in winter conditions exudes between fingers when squeezed in fist	—	0-55

* Loadbearing capacity depends on degree of wetness.
See BS CP 101 *Foundations and substructures* for full details of bearing capacity.

Table II Classification of sulphate soil conditions affecting concrete, and recommended precautionary measures

Classification of soil conditions			Precautionary measures		
Class	Sulphur trioxide in ground water (Parts SO₃ per 100 000)	Sulphur trioxide in clay (per cent SO₃)	Precast concrete products	Cast in-situ concrete Buried concrete surrounded by clay	Concrete exposed to one-sided water pressure, or concrete of thin section
1	Less than 30	Less than 0·2	No special measures.	No special measures, except that the use of lean concretes (eg 1:7, or leaner, ballast concrete) is inadvisable if SO₃ in water exceeds about 20 parts per 100 000. Where the latter is the case, Portland cement mixes not leaner than 1:2:4, or, if special precautions are desired, pozzolanic cement or sulphate-resisting Portland cement mixes not leaner than 1:2:4 should be used.	No special measures, except that when SO₃ in water is above 20 parts per 100 000, special care should be taken to ensure the use of high quality Portland cement concrete, if necessary 1:1½:3 mixes; alternatively, pozzolanic cements or sulphate-resisting Portland cement may be used in mixes not leaner than 1:2:4.
2	30 to 100	0·2 to 0·5	Rich Portland cement concretes (eg 1:1½:3) are not likely to suffer seriously, except over a very long period of years. Alternatively, either pozzolanic, sulphate-resisting Portland or supersulphate cement should be used.	Rich Portland cement concretes (eg 1:1½:3) are not likely to suffer seriously over a short period of years, provided that care is taken to ensure that a very dense and homogeneous mass is obtained. For most work, and particularly if the predominant salts are magnesium or sodium sulphates, concrete made with either pozzolanic cement, sulphate-resisting Portland cement, or supersulphate cement (1:2:4) is advisable. See Note 1.	The use of Portland cement concrete is not advisable. Pozzolanic cement or sulphate-resisting Portland cement or, preferably, supersulphate cement is recommended.
3	Above 100	Above 0·5	The densest Portland cement concrete is not likely to suffer seriously over periods up to, say, 10-20 years, unless conditions are very severe. Alternatively, high alumina or supersulphate cement concretes should be used.	The use of supersulphate cement concretes is recommended.	The use of supersulphate cement concretes is recommended

Notes
1 Where 1:2:4 concrete is mentioned, other mixes of equivalent weight ratio of cement to total aggregate, but with somewhat increased ratio of sand to coarse aggregates (eg 1:2½:3½, or even 1:2½:3½), may be used, sometimes with advantage. It may be necessary when using supersulphate cement to employ mixes somewhat richer than 1:2:4 in order to obtain adequate workability.
2 Adequate assurance should be obtained that cements claimed to be sulphate-resisting Portland cements have, in fact, a high resistance to sulphates. Where the whole of the work, under adverse conditions, cannot be done with resistant cements, protection should be given either by casing with a layer of resistant-cement concrete or by coating with bituminous materials.

1a and **1b** show typical sections through concrete-paved banks and slopes.

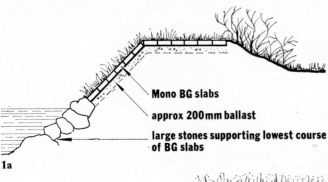

Mono BG slabs

approx 200mm ballast

large stones supporting lowest course of BG slabs

1a

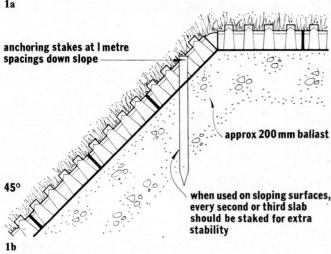

anchoring stakes at 1 metre spacings down slope

approx 200mm ballast

45°

when used on sloping surfaces, every second or third slab should be staked for extra stability

1b

3 Retaining walls

3.01 Design procedure

Soils cannot stand more steeply than their natural angle of repose (see Table I), so that it is often necessary to construct a retaining wall. After an analysis of subgrade conditions and soil profile has been done, the normal design process is:
1 estimate the forces exerted by the material to be restrained on the back of the wall;
2 determine profile of retaining wall and base so that the structure will be stable and not fail by overturning;
3 analyse the wall itself in terms of structural stability;
4 check the bearing pressure under the base, and check whether base is deep enough to prevent sliding;
5 design structural elements;
6 decide drainage method in backfill;
7 allow for movement and settlement;
8 decide on finishes to wall (see Information sheets 9 to 16).
Steps 1 to 5 may require the services of a structural consultant, but for relatively simple cases see *AJ Handbook of building structure*, Information sheet Foundations 8, pages 155-156.

3.02 Retaining wall types

Retaining walls, **2**, **3** can be classified into the following groups (see next page).

1a *Embankment, crown of which forms negotiable traffic surface; stabilising slabs are supported by large stones or rocks along base. Construction must begin at lowest end.*
1b *Alternative anchoring method; stout wooden stakes driven one metre into bank, two or three slab widths apart.*
2a, b *Low concrete retaining walls at Liestal, Switzerland; and Vienna. In British climate, rain streaking of exposed surfaces would be a problem.*

2a

2b

3a

3b

3c

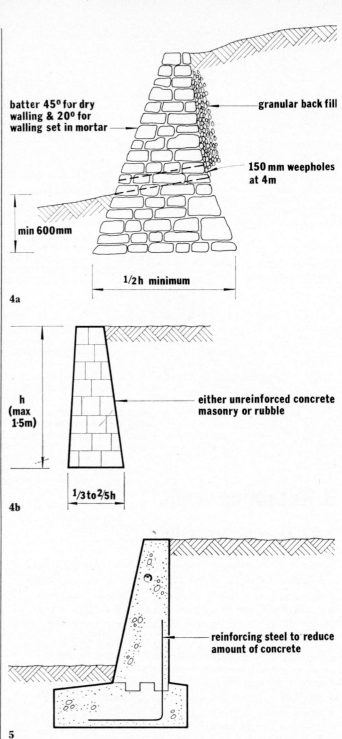

batter 45° for dry walling & 20° for walling set in mortar

granular back fill

150 mm weepholes at 4m

min 600mm

1/2h minimum

4a

h (max 1·5m)

either unreinforced concrete masonry or rubble

1/3 to 2/5h

4b

reinforcing steel to reduce amount of concrete

5

3a *Low retaining wall formed of U-shaped precast concrete blocks, which are also used for steps linking two levels;* **b**, *reinforced concrete retaining wall in West* Germany; **c** *crib wall on M6 motorway.*
4a *One form of gravity retaining wall;* **b**, *gravity wall dimensions.*
5 *Semi-gravity wall.*

Gravity walls, 4

These are walls which depend for their stability on dead weight; they are so designed that there is no tensile stress in any part of the structure, and are usually built of unreinforced concrete, masonry or rubble. Gravity walls are primarily trapezoid in section, often with a base which projects beyond the face and/or back of the wall. For unreinforced concrete the width at the top of the wall should be a minimum of 200 mm to facilitate placing and compaction, and the width of the base is normally between one-third and two-fifths the height of the wall.

In terms of economics, gravity walls are only suitable for heights not exceeding about 1·5 m, and where lateral pressures are low.

Semi-gravity walls, 5

These are similar in all respects to gravity walls, except that a small amount of reinforcing steel is incorporated to reduce the amount of concrete and to minimise the possibility of cracking (due either to temperature changes or to shrinkage).

Cantilever walls, 6

Cantilever walls, which are usually in the form of an inverted T, or of an L, are economical up to a height of 7 to 9 m. The material used for construction is almost invariably reinforced concrete; for low walls (up to about 3·5 m high) either standard precast units or blocks may be used, reinforced.

Depending on design requirements, the foot of the

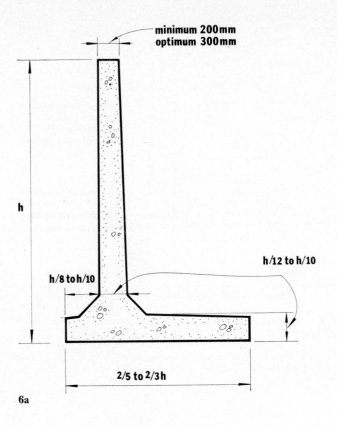

6a

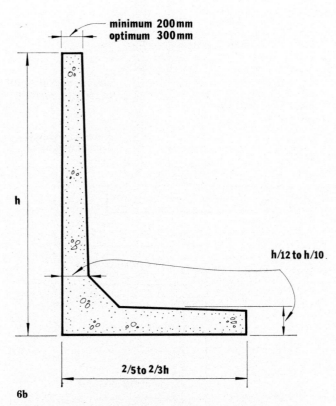

6b

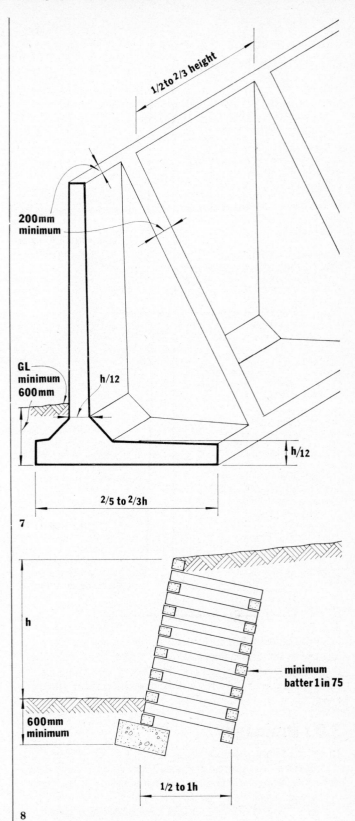

7

8

cantilever may face either into or out of the wall, or both. With the foot turned into the wall so that it is under the material supported, the design takes advantage of the pressure of the material above so that the dead weight of the wall is increased. With the foot turned out from the wall, the main advantage is simplicity of construction, although some form of heel is often necessary for stability.

Counterfort walls, 7
The normal form for the counterfort wall is that both the base slab and the face of the wall span horizontally between vertical supports. For walls up to 10 m high, the counterforts can be spaced as far apart as two-thirds of the height of the wall. For walls higher than this the spacing may be reduced to less than half the height. Because of

construction complications counterforts should never be spaced closer than 2·5 m.

Crib walls, 8
A wide range of crib walling units is available and manufacturers' catalogues should be consulted before making any design decisions. Crib walls are usually constructed with a batter of 1:75 and the base width or depth usually ranges from 50 to 100 per cent of the wall

6a, b *Guide to proportions of inverted T, and L, reinforced concrete cantilever walls.*
7, 8 *Guide to proportions of counterfort and crib walls.*

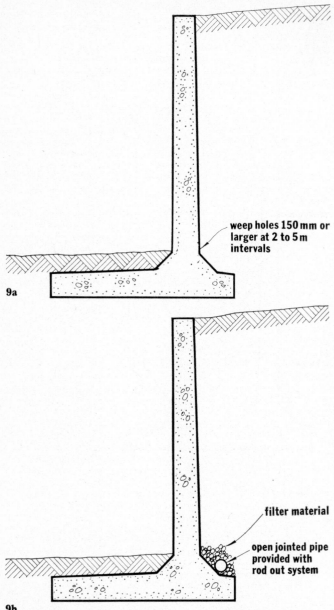

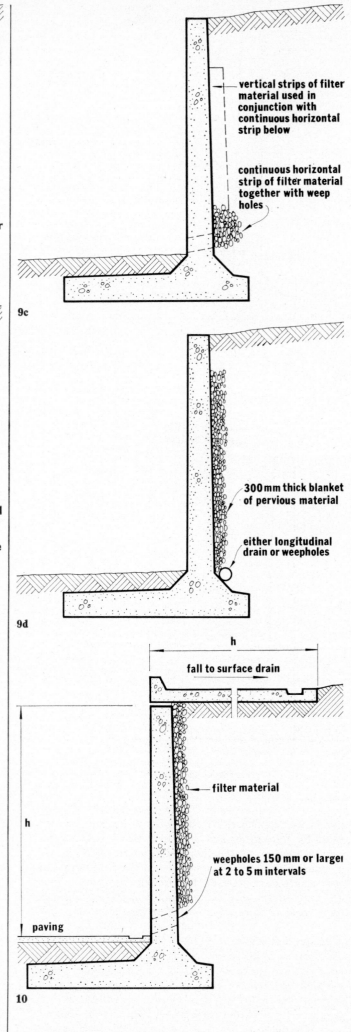

9a weep holes 150 mm or larger at 2 to 5 m intervals

9b filter material / open jointed pipe provided with rod out system

9c vertical strips of filter material used in conjunction with continuous horizontal strip below / continuous horizontal strip of filter material together with weep holes

9d 300 mm thick blanket of pervious material / either longitudinal drain or weepholes

10 fall to surface drain / h / h / paving / filter material / weepholes 150 mm or larger at 2 to 5 m intervals

height. The open box forms making up the box walls are filled with stone or soil to make an economic wall which has the added advantage that it can be planted with climbing shrubs and the like. A crib wall is essentially of the gravity type.

3.03 Drainage

In order to limit the possible build-up of water pressure, it is essential that proper drainage be provided. This can be done by placing suitable drainage material behind the back of the wall (clean sand; gravel; gravel and sand containing less than about 5 per cent of very fine sand, silt or clay particles); and in addition there should be a drainage system to carry the water away. The type of drainage system depends very much on the nature of the backfilling material: guidelines are given in **9a** to **9d**.

Where there is likely to be extensive run-off of surface water behind the wall, it should be intercepted by a continuous drain behind the wall. The drain should be spaced at least the height of the wall behind the retaining wall—see **10**. In addition, water from weepholes in the retaining wall should be collected by a drain at the foot of the wall to prevent the soil under the base being softened.

9a *Pervious backfill and weepholes;* **9b**, *pervious backfill and continuous drain.* **9c** *Semi-pervious backfill and* *weepholes.* **9d** *Fine-grained backfill and continuous drain.* **10** *A method of dealing with surface water run off.*

Information sheet 4
Fences and barriers

1 Concrete fences

1.01 Types

Precast concrete fencing posts are available for carrying almost every type of fencing: they provide durable support which is free from the destructive action of moisture in the ground. There are many different types of fencing and, consequently, there are many appropriate types of concrete fence posts and struts; it is, therefore, strongly recommended that reference be made to British Standard Specification 1722: Fences.

BS 1722 is published in 11 parts—one for each type of fencing in common use. The following list gives the titles of seven of these parts which incorporate concrete posts. Each part specifies the particular type of concrete posts to be used, the method of manufacture and requirements as to erection. See tables I to III.

BS 1722: 1963 Part 1 Chain link fences
            Part 2 Woven wire fences
            Part 3 Strained wire fences
            Part 4 Cleft chestnut pale fences
            Part 5 Close boarded fences including oak
                   pale fences
            Part 6 Wooden palisade fences
            Part 10 Anti-intruder chain link fences

The following list of the more usual concrete fence types (not all of which are covered by the above British Standards) is taken from Data sheet D.16.1 (December 1973) of the BPCF*.

Posts for ornamental chains
These can be used for the frontages of domestic or industrial premises, and in public parks. The posts are in the form of short 'bollards' with domed or differently shaped tops. Metal fixings to which ornamental chains can be attached are cast in two opposite sides near the top. They are available in a range of surface finishes—generally exposed aggregate.
Typical heights: 825 mm and 900 mm.

Posts and rails
Post and rail fences are used for the boundaries to property, parks, cattle enclosures, roadsides etc. The posts are generally square in section (often with chamfered edges) but octagonal sections are also available. The tops are generally rounded one way but can be square or pointed.

*British Precast Concrete Federation, 60 Charles Street, Leicester, LE1 1FB.

Table I Concrete posts for chain link fences

Fence type	Number of holes for wires	Intermediate posts		Straining posts		Struts*	
		Length	Base dimensions	Length	Section	Length approx	Section
GLC 90, GLC 90A PLC 90, PLC 90A and B	3	1·60 m	100 × 100† mm	1·60 m		1·50 m	100 × 75 mm
GLC 120, GLC 120A PLC 120, PLC 120A	3	1·87 m	125 × 125 mm	1·87 m		1·82 m	100 × 75 mm
GLC 140, GLC 140A–D PLC 140, PLC 140A–G	3	2·07 m	125 × 125 mm	2·07 m	125 × 125 mm	1·98 m	100 × 75 mm
GLC 180, GLC 180A–B PLC 180, PLC 180A–B	3	2·63 m	125 × 125‡ mm	2·63 m		2·59 m	100 × 85 mm
GLC 213, GLC 213A PLC 213, PLC 213A	6	3·04 m§	125 × 125‡ mm	3·04 m		2·59 m	100 × 85 mm

* These lengths are suitable for struts fixed at an angle of 45° on level ground. If site conditions make the use of struts of these lengths unsuitable, the length shall be agreed between the purchaser and the supplier.
† Base dimensions of intermediate posts for types 90 are acceptable up to 115 mm.
‡ Bases of posts with standard taper are acceptable up to 145 mm × 145 mm.
§ Measured along the centre line. The top 0·43 m length of the intermediate posts and the top 0·45 m length of the straining posts shall be cranked at an angle of 40–45° to the vertical to bring the total vertical height to approximately 3·1 m. The direction of the crank shall be inwards unless otherwise specified at time of ordering.

Table II Concrete posts for woven wire fences

Fence types	Number of horizontal wires	Intermediate posts		Straining posts		Struts*	
		Length	Base dimensions	Length	Section	Length approx	Section
C5/75/30C	5	1·45 m	100 × 100 mm	1·60 m		1·40 m	
B8/80/15C, C8/80/15C, B8/80/30C	8	1·50 m	100 × 100 mm	1·70 m	125 × 125 mm	1·45 m	100 × 75 mm
B6/90/30C, C6/90/30C	6	1·60 m	100 × 100 mm	1·75 m		1·50 m	
B8/115/30C, C8/115/30C	8	1·80 m	125 × 125 mm	2·00 m		1·75 m	

* These lengths are suitable for struts fixed at an angle of 45° on level ground. If site conditions make the use of struts of these lengths unsuitable the length shall be agreed between the purchaser and the supplier.

Table III Concrete posts for strained wire fences

Fence types	Number of holes for wires	Intermediate posts		Straining posts		Struts*	
		Length	Base	Length	Section	Length approx	Section
SC 85	3	1·50 m	100 × 100 mm	1·50 m	100 × 100 mm	1·45 m	75 × 75 mm
SC 100A	5	1·67 m	100 × 100 mm	1·82 m	125 × 125 mm	1·50 m	100 × 75 mm
SC 100B	6	1·67 m	100 × 100 mm	1·82 m	125 × 125 mm	1·50 m	100 × 75 mm
SC 120	6	1·90 m	125 × 125 mm	2·02 m	125 × 125 mm	1·80 m	100 × 75 mm
SC 140A SC 140B	8	2·70 m	125 × 125 mm	2·22 m	125 × 125 mm	1·98 m	100 × 75 mm

* These lengths are suitable for struts fixed at an angle of 45° on level ground. If site conditions make the use of struts of these lengths unsuitable, the length shall be agreed between the purchaser and the supplier.

The posts are mortised to receive one, two or three reinforced concrete, **1a**, or round galvanised steel, **1b**, rails. Typical heights: 1·5 m; 1·65 m; 1·8 m.

Picket fences

These white concrete fences are often used for boundaries to parks or house properties. They consist of square, pointed-topped posts mortised to receive the top and bottom rails of precast panels of open pales, also with pointed tops: the rails can be grouted in to the posts for permanency. These fences are adaptable for sloping sites, in which case the posts have their mortises higher on one side than the other, so that each panel can be set higher or lower than its neighbour.
Typical heights: 900 mm, 975 mm and 1·2 m.

Palisade fences

Palisade fences resemble picket fences but are generally high and the pales closely spaced to prevent intrusion on property, **2**. They are therefore suitable for factory boundaries, public buildings, sports fields and the like. The square posts are reinforced for greater strength—as are the pales.
Typical heights: 1·2 m, 1·5 m, 1·8 m and 2·44 m.

Post and panel fencing

Unlike previous types, this form of fencing provides visual privacy as well as physical protection. Its solidity makes it a very intrusive element in the landscape, however. There are two types.
● Posts and gravel boards for close boarded timber fences. The precast concrete posts have grooved sides to receive complete bays of vertical or horizontal timber weatherboarding: the posts are also recessed at ground level to receive a concrete gravel board.
The posts and gravel boards can be of plain, smooth, natural coloured concrete or they are obtainable with their outer surface finished with exposed aggregate in a wide range of colours and textures, as described for 'ranch' type fencing.

Typical heights: 1·2 m, 1·8 m.
Gravel boards: 300 mm high.
● Concrete post and panel walling.
The posts are as last described, with the sides grooved, but the panels are of precast, reinforced concrete. The horizontal panels are available in a number of different designs. They can be supplied butt-jointed or rebated on their lower edges to produce a ship-lap joint and weatherboard profile. They can also be supplied with a 'reeded' surface or with textured exposed aggregate. The topmost of the horizontal panels can be supplied in fretted, openwork or trellis designs or can have a contrasting pattern of vertical reeding above the horizontal reeded panels.
Typical heights: 900 mm, 1·2 m, 1·5 m, 1·8 m, 2·15 m and 2·44 m.

'Ranch' fencing

This fencing is suitable as a boundary to domestic, agricultural or industrial premises. It consists of posts supporting two, three or four wide horizontal panels with spaces between, **4**. The outer faces of the posts and panels are faced with rough-textured self cleaning exposed granite aggregate chippings; the reverse faces of the units are smooth concrete. These fences are particularly for sloping sites.
Typical heights: 900 mm, 1·38 m and 1·8 m.

Mini walling

This is a low boundary enclosure and consists of short posts with a specially shaped recess in the top to receive the end of a single wide panel (adjacent panels butt together). This fencing has the same range of textured finishes as ranch fencing.
Height: 450 mm.

Plastic coated concrete fencing

This type of fence is sheathed in a green, grey or black plastic coating. The slim, round reinforced concrete posts and struts are encased in unplasticised pvc, which extends

1a

1b

3

2

4

below ground level; the tops of the posts are finished off with plastic caps. The chain link fencing which it supports is also coated with plastic, as are the attachments for it on the posts; matching gate posts are also available. This type of fence is particularly suitable for enclosing tennis courts. Heights: 900 mm, 1·22 m, 1·38 m, 1·8 m, 2·4 m and 2·8 m.

Anti-intruder fencing
Although they are described in part 10 of the British Standard, some manufacturers produce anti-intruder fence posts to their own specification. The stout reinforced concrete posts, generally supporting chain link fencing, have cranked tops sloping inward, carrying several strands of barbed wire to deter intruders. This fencing is particularly used for the boundaries of industrial premises.
Typical heights: 2·55 m and 2·95 m.

1.02 Materials

Portland cement (ordinary or rapid hardening) should conform to BS 12. BS 1722 permits the use of portland blastfurnace cement—BS 146—or cements complying with used for the boundaries of industrial premises.
The aggregates should conform to BS 1047 or BS 882.
The sizes of the reinforcement for the posts are given in

BS 1722. The bars should be four in number, in the form of a prefabricated cage with stirrups. Minimum cover to steel 15·8 mm. The British Standard requires that the reinforcing gauge shall be controlled in position to ensure that the minimum cover is achieved.
The concrete should attain a minimum compressive strength of 25·8 N/mm^2 (3750 lbf/in^2) at 28 days or 17·2 N/mm^2 (2500 lbf/in^2) at seven days.
The water/cement ratio should be as low as possible consistent with achieving full compaction by vibration. Control of quality may be exercised by works control test cubes taken from the concrete delivered at the place of moulding. Alternatively, the purchaser may specify proof loading tests on the finished posts. The method of test and the standard to be achieved are specified in the British Standards. The purchaser may also check the cover of the reinforcement by visual examination. For this test the purchaser may break one post out of each consignment (with a maximum of one in 100).
Erection is fully dealt with in BS 1722 and it is essential to comply with these directions.

1a, b *Post and rail. fences,*   **2, 3** *Palisade fence.*
the first with concrete rails;   **4** *Ranch walling.*
and the second with metal.

Information sheet 5
Walls and screens

1 Solid and screen walls

Solid concrete walls can be formed either in situ (unlikely, except in the case of retaining walls; see Information sheet 3 for details on the latter), or by using blocks.

Screen walls will almost invariably be formed from blocks. In their simplest form, they may consist of standard units laid so that there are gaps between the blocks, or standard cored units placed on their sides. There is, however, a growing demand for special pierced blocks, **1**, which can be manufactured on standard block-making machines using purpose-made moulds.

2 Block types

2.01 Classification

BS 2028, 1364: 1968 classifies concrete blocks into three types: A, B and C (see table I). The distinction between the three types is based on block density, calculated by dividing the weight of the block by the overall volume (including holes and cavities).

Type A is intended for general use in building, including use below ground level dpc.

Type B is also intended for general use in building, with qualifications; the latter however apply to buildings rather than landscaping uses and are therefore not discussed here. Type B blocks (hollow or solid blocks of cement, sand and lightweight aggregate or graded wood particles, or similarly produced units) are suitable for walling.

Type C blocks are primarily for internal non-loadbearing walls in buildings, such as partitions and panels in framed construction.

2.02 Forms of blocks

In BS 2028, 1364: 1968 a block is defined as a walling unit of length, width or height greater than that specified for a brick. To avoid confusion with slabs or panels the definition of a block further states that the height of a block shall not exceed either its length or six times its thickness. Concrete blocks can be solid, hollow or cellular, as described below.

Solid blocks
These are essentially voidless, but they can have end grooves, finger holes or other small cavities to improve handling and reduce weight, provided the total volume of such cavities does not exceed 25 per cent of gross volume of block.

Table I Summary of main requirements of BS 2028, 1364 (published by courtesy of BSI)

Block type	Materials and methods of manufacture	Density of block	Minimum average compressive strength N/mm²*	Strength—lowest individual block N/mm²*	Maximum permitted drying shrinkage (per cent)	Wetting expansion
A	Any combination of materials and methods of manufacture may be used provided resulting blocks comply with the specification	Not less than 1500 kg/m³	3·5 7·0 10·5 14·0 21·0 28·0 35·0	2·8 5·6 8·4 11·2 16·8 22·4 28·0	0·05 0·05 0·06 0·06 0·06 0·06 0·06	This applies only to blocks made with clinker aggregates and is to be not more than 0·02 per cent in excess of drying shrinkage value
B	As above	Less than 1500 kg/m³ but more than 625 kg/m³	2·8 7·0	2·25 5·6	0·07 0·08	
		Less than 625 kg/m³	2·8	2·25	0·09	
C	As above	Less than 1500 kg/m³ but more than 625 kg/m³	Transverse breaking load is specified (varies with size of block)		0·08	
		Less than 625 kg/m³			0·09	

*N/mm²—Newtons per sq mm

Hollow block

These blocks have more obvious cavities which pass right through the block; total volume of cavities is restricted to not more than 50 per cent of the gross volume of the block.

Cellular blocks

These are a special type of hollow block in which the cavities are closed at one end. As in a hollow block, total volume of cavities must not exceed 50 per cent of the gross volume of the block.

Blocks with special faces

These may be blocks with profiled faces or blocks with a special face applied to a backing of dense or lightweight concrete as an integral part of the manufacture, or they may be blocks to which a special face is applied subsequent to moulding. They are deemed to comply with BS 2028, 1364: 1968 provided they comply with the test requirement appropriate to their type.

3 Dimensions

The main considerations for deciding upon the size of blockwork should include the following.

3.01 Handling and laying

Blocks should not be too large to be laid by one person. The heaviest solid block that can be lifted and positioned with one hand is about 10 kg; with two hands (provided there is an adequate grip) 30 kg. To facilitate singlehanded laying, efforts should therefore be made to use blocks of around 10 kg; reducing weight below this figure will be of little value and could lead to unproductive effort by the blocklayer.

The shape of a block also appreciably influences the ease with which it can be handled; a block of about 200 mm thickness can be grasped easily with one hand, unlike thicker blocks. Blocks over 100 mm thick are usually hollow to reduce weight and make them easier to handle.

Mortar

The fewer the joints the less mortar will be required in laying, and the less labour will be required for pointing and tooling of joints. Relative costs of mortar and blocks should be considered: in facing work the mortar may represent a smaller proportion of overall cost of wall and may not be critical.

Table II Dimensions of blocks incorporated in the 1970 amendment to BS 2028, 1364: 1968 (published by courtesy of BSI)

Block type	Length × Height Co-ordinating size (mm)	Work size (mm)	Thickness Work size (mm)
A	400 × 100 400 × 200	390 × 90 390 × 190	75, 90, 100, 140 and 190
	450 × 225	440 × 215	75, 90, 100, 140, 190 and 215
B	400 × 100 400 × 200	390 × 90 390 × 190	75, 90, 100, 140 and 190
	450 × 200 450 × 225 450 × 300 600 × 200 600 × 225	440 × 190 440 × 215 440 × 290 590 × 190 590 × 215	75, 90, 100 140, 190 and 215
C	400 × 200 450 × 200 450 × 225 450 × 300 600 × 200 600 × 225	390 × 190 440 × 190 440 × 215 440 × 290 590 × 190 590 × 215	60 and 75

Note Blocks of work size 448 mm × 219 mm × 51, 64, 76, 102, 152 or 219 mm thick, and 397 mm × 194 mm × 75, 92, 102, 143 or 194 mm thick will be produced as long as they are required.
If blocks of entirely non-standard dimensions or design are required the limits of size or the design shall be agreed. Such blocks shall then be deemed to comply with this standard provided they comply with the other requirements.

3.02 Appearance

There has hitherto been a tendency to limit the sizes of facing blocks to those of bricks, but large blocks need not be unsightly if due consideration is given to the proportions of wall panels, treatment of joints, and bonding pattern, **2**. Larger dimensions can be used to advantage by emphasising either the horizontal or the vertical joints.

An amendment to BS 2038: 1968 indicates a wide range of block sizes (which tends to defeat the whole object of dimensional co-ordination—that of variety reduction); see table II.

1 *Miscellaneous examples of screen walls formed of pierced concrete blocks. Blocks used in* **b, c** *and* **d** *are commercially available screen wall units— manufacturers' catalogues should be consulted.*

2 *Bond patterns:* **a** *running bond (400 × 200 mm);* **b** *stack bond (200 × 200 mm);* **c** *vertical stack bond (400 × 200 mm);* **d** *stack bond (400 × 200 mm);* **e** *coursed ashlar (400 × 200 mm and 400 × 100 mm);* **f** *coursed ashlar-alternate courses (400 × 200 mm and 400 × 100 mm);* **g** *basket weave pattern (400 × 200 mm);* **h** *patterned ashlar (400 × 200 mm and 200 × 200 mm).*

1a

1b

1c

1d

1e

2a

2b

2c

2d

2e

2f

2g

2h

3.03 Standard block sizes

Manufacturers should be consulted prior to design and detailing, to determine the range of blocks available, in order to avoid a multiplicity of 'specials' or the expensive alternative of cut blocks (particularly with dense concrete). Most manufacturers supply half-length blocks in addition to the standard unit; and many will also supply standard quoin blocks, bond beam and lintel blocks, **3**.

Concrete blockwork masonry walls should be laid out to make maximum use of these standard full and half-length units—this applies to all dimensions, such as overall length and height of walls; width and height of any openings; and width and height of panels between gates and corners. Where thickness of wall is greater or less than the length of a half unit, a special-length unit is required at each corner in each course.

4 Properties

4.01 Dimensional changes

All masonry materials (stone, clay brick, concrete brick and concrete block) undergo changes of dimension with wetting and drying and with changes of temperature.

The expansion in volume of concrete products caused by increasing moisture content—originally called 'moisture movement' but now more fittingly termed 'wetting expansion' in the 1968 British Standard—is usually slightly less than the value of drying shrinkage. It gives rise to compressive stresses if the movement is restrained, but is unlikely to be a significant factor affecting the cracking of walls containing concrete blocks.

With concrete products additional non-reversible shrinkage takes place as a result of chemical changes within the composition of the concrete (associated mainly with the processes of carbonation, hydration and curing).

These dimensional changes are small, but taken together with those caused by other factors, they are sufficiently important to require special attention to the problem of detailing, if risk of cracking is to be minimised—see paragraph 5.04 for guidance.

4.02 Strength

BS 2028, 1364: 1968 specifies seven grades of strength for type A blocks and two grades for type B, with average compressive strengths up to 35 N/mm² (see table I). Provision is made for other strengths to be agreed between the purchaser and the supplier. In this context, strength is the average compressive strength of a test sample of 10 blocks; the compressive strength of the weakest individual block in the test sample must not be less than 80 per cent of the average value. In this way control is exercised on the quality of the blocks in the consignment as a whole.

Type C blocks (non-loadbearing) are subjected to a transverse breaking load test, and the specified values vary with size and thickness of blocks.

Solid, hollow and cellular loadbearing blocks are tested for compressive strength in the same manner (compressive strength is obtained by dividing crushing load by the overall cross-sectional area of the block); thus there is no difference in the load-carrying capacity of solid blocks and of hollow or cellular blocks of equal compressive strength.

4.03 Durability

Concrete is an inherently durable material, as is concrete blockwork. In normal circumstances no problems should be encountered, provided that efficient dpcs, flashings, weathering and coping systems are provided (see

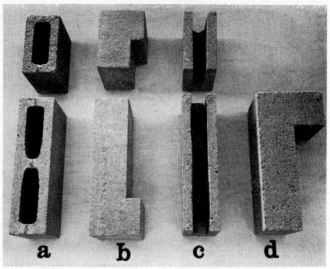

3 *shows a kit of parts which manufacturers should be able to supply as standard items:* **a** *full and half length blocks;* **b** *full and half length cavity* closers; **c** *full length bond beam block and half length lintel block;* **d** *full length quoin block.*

paragraphs 6.01 to 6.03). Where atmospheric pollution is particularly strong (in certain industrial areas for example) and where particularly adverse weather conditions are found, facing blocks should have an average compressive strength of not less than 7 N/mm². Open textured blocks are not, as might at first be thought, susceptible to frost damage, because their open structures permit water to move freely within the composition of the block without building up disruptive forces caused by the formation of ice in frosty weather.

5 Design considerations

5.01 Volume stability

A characteristic of Portland cement is its tendency to decrease in volume as the moisture content decreases. The drying out of a concrete blockwork wall will therefore produce a tendency for it to shrink which, if the wall is restrained, will result in tensile stresses. If these stresses exceed the tensile strength of the wall, cracking will occur. Unit shrinkage of a concrete block is dependent on:
1 physical properties of the aggregate;
2 chemical and physical properties of the cement;
3 gradation of the aggregate;
4 elastic properties of the aggregate;
5 aggregate/cement ratio;
6 method of curing the unit;
7 ultimate tensile strength of the unit or mortar;
8 elastic properties of the unit; and
9 design of the unit.
As the many investigations continue, the following recommendations are a practical and economical approach to the minimising of volume changes caused by moisture:
1 choose a concrete block with favourable dimensional stability;
2 keep blocks as dry as possible at all stages, particularly during site storage as well as during laying operations;
3 stack blocks on planks off the ground to allow air circulation and to avoid absorption of moisture from the ground;
4 cover storage piles;
5 do *not* dampen blocks prior to laying; and
6 cover work in progress whenever work is suspended.

5.02 Mortars

The type of mortar mix used will depend upon such factors as: type of wall; type of block; location; degree of exposure to which it is subject and speed of construction required. In general blocks should be laid in the *weaker* mortars to confine cracks to joints. Under normal conditions mortars of only moderate strength are adequate and are to be preferred (see table III).

5.03 Joints

Thickness
Uniform thin joints are desirable since the thinner the joints the stronger will be the wall. Such joints are, however, extremely difficult to obtain, since they require a very high standard of workmanship, and uniformity in the size and shape of the units. The thickness of 10 mm which is generally used is normally satisfactory and has the advantage that blocklayers are familiar with working to this thickness.
In facing blockwork, recessed joints are to be preferred because they give a cleaner appearance, particularly with open textured units. They also help to obscure any cracks in the joints.

Resistance to water penetration
To avoid water penetration give careful attention to joints, which should be designed to shed rain away from the wall interior, and constructed to be as well-filled as possible. Otherwise penetration may occur by capillarity either through cracks within the mortar or through the plane of contact between the mortar and the blocks.
Weathered, concave, and vee joints are best for shedding water, 4. In addition to their weathering profile, their formation requires pressure sufficient to compress the mortar and create a firm bond between the mortar and the units at the face of the wall. Raked and struck joints form ledges on which water can collect and penetrate into the blockwork.
Bed joints must be full and not excessively furrowed. Tool finishing of the joint should be delayed until the mortar has stiffened. Where pointing is used it should be at least 10 mm deep and in a mortar of colour to suit that of the blocks.

Bedding
Strength of joint depends on area of bedding. The greater the bedding area, the greater the strength of the wall. 'Face-shell' bedding, for instance, produces a wall which is weaker than full bedding; and an important factor in the reduction of the strength of cored-block walls is the smaller bedding area at the joints.

Dry bonded blockwork
This falls broadly into two categories:
1 units designed to interlock with each other, which may be likened to a 'tongue and groove' arrangement;
2 units designed to permit the insertion of some device which locks the blocks and stabilises the wall.
The chief advantage of dry-bonded blockwork is that by the elimination of mortar it speeds up building considerably and may improve the strength and the appearance of the wall.

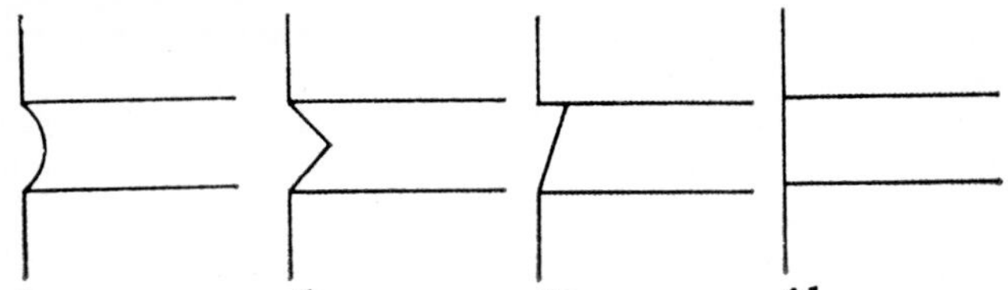

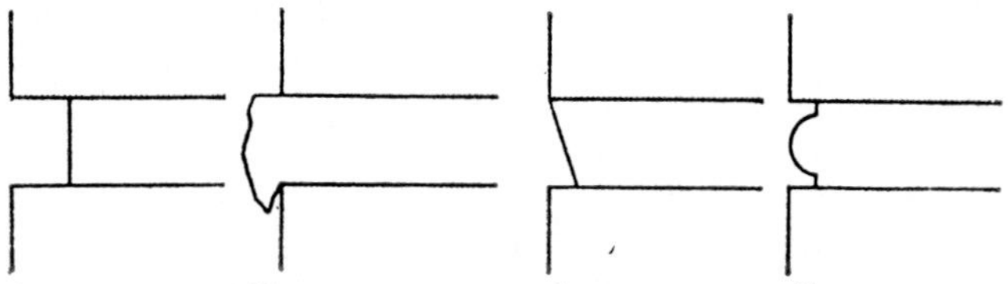

4 *Joint types;* **a** *concave;* **b** *vee-joint;* **c** *weathered;* **d** *flush;* **e** *raked;* **f** *squeezed;* **g** *struck;* **h** *beaded.* **4a** *to* **d** *are recommended as being relatively weather tight,* *while* **e** *to* **h** *are not recommended as the horizontal ledges collect water which tends to cause moisture penetration.*

Conclusion
It cannot be over-emphasised that joints are one of the most critical features of blockwork. Untidy and careless jointing will ruin joint performance and the appearance of a wall. Often expensive blockwork is ruined because of inadequate attention to joints. A sample panel of the blockwork should be erected and approved and serve as a yardstick for the whole project. The panel may highlight otherwise unforeseen jointing problems.

5.04 Controlling of cracking

Control joints (usually vertical) should be located where cracking would be most likely to occur in long straight walls, and at other points of potential excess tensile stress such as abrupt changes in wall thickness, 5; at openings; at junctions with columns and intersecting walls, 6; and at places with major changes in wall height, 7.
Cracking is more likely to occur if the length of a panel exceeds about 1½ to 2 times its height; and shape appears to be more critical than size. The shorter the panel, the better. The most successful technique is to divide the blockwork into rectangular panels without any openings and with straight vertical joints between adjacent panels, thus eliminating the problems of concentration of stresses at

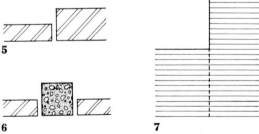

5 *to* **7** *show potential sources of cracking in masonry walls, which may require control joints.* **5** *changes in wall thickness;* **6** *junctions with columns and intersecting walls;* **7** *major changes in wall height.*

Table III Recommended mortar mixes (proportions by volume)

Direction of change in properties	Mortar designation	Hydraulic lime: sand	Cement: lime: sand	Masonry cement: sand	Cement: sand with plasticiser
Increasing strength but decreasing ability to accommodate movements due to settlement, shrinkage, etc	(i)	—	1: 0-¼:3	1:2-2½	1:2½-3
	(ii)	—	1:½:3½-4½	1:3-3½	1:3½-4
	(iii)	—	1:1:5-6	1:4-5	1:5-6
	(iv)	1:2-3	1:2:8-9	1:5½-6½	1:7-8
	(v)	1:3	1:3:10-12	1:6½-7	1:8
Changing characteristics within any one mortar designation	→ Increasing resistance to damage by freezing after hardening				
	← Improvement in bond and consequent resistance to rain penetration				

internal angles. A straight vertical joint should separate panels of blockwork from each other, and a block panel from its adjoining wall element. Any finishes applied to the blockwork should form a clear break at the panel joints, and the visual effect of these joints should be carefully considered and integrated into the overall design of the wall. A control joint, **8**, may be formed by first laying the blocks in the wall in the usual manner, after which the mortar at the control joint is raked to a depth of not less than 12 mm. The control joint is then caulked with an elastic compound (the edges of the block in the control joint may have to be primed before caulking to prevent the dry masonry from absorbing oils from the compounds).

Reinforcement. If it is impracticable to subdivide walls into panels, use joint reinforcement to control cracking by distributing stresses more uniformly throughout the wall. Use of prefabricated joint reinforcement has proved to be effective; while it will not necessarily eliminate all cracks, those that occur will be extremely small and scarcely noticeable. Joint reinforcement also adds strength to walls to assist in cases of unpredictable settlement or lateral loads. The extent of joint reinforcement required depends on the characteristics of the masonry unit, length and height of wall, location of openings, and spacing of control joints (seek an engineer's advice in complex cases).

Joint reinforcement should be placed at least 20 mm from the face of the mortar so that there is adequate cover, and to reduce the risk of rust staining the blockwork.

Additional steel is recommended at the top of walls, particularly those not restrained against shrinkage along the top edge. Cracks usually originate at the top of the wall, or at lintel and sill levels of openings; reinforcement is most effective at these places.

When bond beams are used joint reinforcement seldom is placed within 600 mm of a bond beam, and usually is omitted altogether if bond beams are spaced at 1200 mm or less. It is general practice to install joint reinforcement at 400 mm vertical intervals, **9**, although a 200 mm spacing can be used to advantage when the widest possible spacing between control joints is desired. With 400 mm spacing, additional steel should be placed above and below openings and at other points of critical stress not relieved by control joints.

Almost without exception horizontal reinforcement should be discontinuous at control joints. Exceptions to this rule should be decided by the designer and not in the field. However, the stability of the block wall on one side of the joint may need to be increased by the use of non-corroding dowel bars placed across the joints in every other course.

5.05 Wall thickness

The following thicknesses are required for stability of screen block walls as a general rule.

Minimum thickness of *reinforced non-loadbearing* walls is 100 mm, and maximum unsupported height or length (or both), 3 m.

Minimum thickness of *unreinforced non-loadbearing* walls is 100 mm, with a maximum unsupported length, height (or both) of 2 m.

Minimum thickness of *unreinforced loadbearing* walls is 150 mm, with a maximum length, height (or both) of 2·7 m; and for *reinforced loadbearing walls* the corresponding figures are minimum thickness 150 mm, and maximum dimensions 3·75 m.

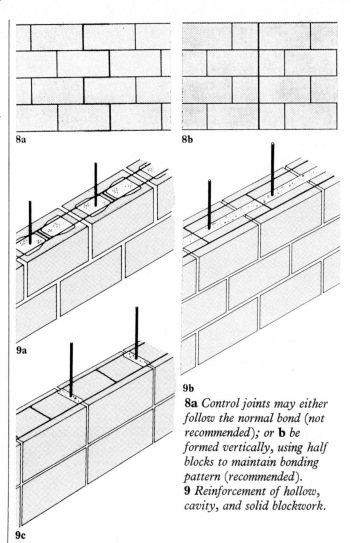

8a *Control joints may either follow the normal bond (not recommended); or **b** be formed vertically, using half blocks to maintain bonding pattern (recommended).*
9 *Reinforcement of hollow, cavity, and solid blockwork.*

6 Design details

6.01 Damp-proof courses

Dpcs should provide a barrier to passage of water. Any course upon which a damp-proof material is to be laid should be carefully flushed up with mortar to form an even bed. Then lay the dpc and protect from injury until the mortar is set. Take care that the mortar will not injure a damp-proof material. Materials suitable for damp-proof courses are described in the following British Standards: BS 743: 1970 Material for damp-proof courses; BS 1097: 1966 Mastic asphalt for building (limestone aggregate); BS 1162, 1410, 1418: 1966 Mastic asphalt for building (natural rock asphalt aggregate); BS 1521: 1965 Waterproof building papers; BS 2870: 1968 Rolled copper and copper alloys. Sheet, strip and foil; BS 1178: 1969 Milled lead sheet and strip for building purposes.

Polythene sheet is also widely used for damp-proof courses in walls; the manufacturers should submit evidence that the material is suitable for the intended purpose.

6.02 Flashings and weatherings

Flashings and weatherings should be selected with due regard to conditions of use, exposure and possible chemical action resulting from contact with other materials.

6.03 Solid copings

Non-metallic copings should conform to the requirements of BS 3798: 1964 Coping units (of clayware, unreinforced cast concrete, unreinforced cast stone, natural stone and slate), or of materials not inferior to such copings.

Information sheet 6
Services and drainage

1 Underground service layouts

1.01 Introduction

The existence of pipeline services beneath hard landscaped surfaces causes problems when the pavings have subsequently to be taken up for maintenance of the buried services. Ideally, therefore, there should be no service pipes of any kind laid under pavings. Unfortunately this is not always possible, or within the landscaper's area of responsibility; and the following paragraphs give advice on minimising the disruptive effects of such buried services on the pavings above.

Services can, for the purposes of this sheet, be grouped into two categories: *supply* (gas and water mains and telephone and electricity cables) and *disposal* (sewers and foul and surface water drains).

1.02 Supply services

Some services may require special safety precautions when they are laid in proximity, such as gas and electrical lines. The width of the resulting channel may be quite wide, but it must be located for ease of access, and it should be specially constructed with pavings which are easy to lift. The simplest and most desirable treatment for gas, water, telephone and electric mains is to place them on one or the other side of the paved route: under the footway or, preferably, under a grassed area. The layout should ensure that the need for cross ducts at frequent intervals under the paving is eliminated, and if necessary duplicate mains on both sides of the paving should be used to avoid the need for cross ducts.

If the layout has been determined and the position and size of the services and their connections is known, they can be laid before any paving is constructed. This, of course, demands a high degree of co-ordination between local and statutory authorities, and may in any case not be possible where a hard paving is needed at an early stage to give access to a difficult site or rapid protection to the paving sub-grade. In this case, it is advisable to leave out narrow cross bays along the line of the proposed service runs to enable the latter to be laid without disturbing other finishes. The cross bays can then be completed when all services have been laid. The main run of services should, however, still be laid under a footpath or verge. As a principle, the trenches for cross-connections should be dug and reinstated as long in advance of the paving as possible. This will give the filled material a better chance of consolidating to the same degree as the surrounding ground; if it does not, the area will be weak and any surface subsequently laid over it will be liable to cracking or subsidence. When the programme of the builder and the statutory authorities cannot be closely combined, future provision for cross-connection services can be made by introducing ducts at frequent intervals, through or under the foundations, before the paving is laid. Hydraulic thrust-boring can also be a practicable and economic proposition. Where this method is used no preliminary provision for services is necessary, as a hole for the pipe or cable can be bored under the finished paving as required. Before thrust-boring can begin care must obviously be taken to ascertain the exact position of any existing mains and services.

1.03 Disposal services

The question of sewerage layouts is much more complex. Every site calls for a different treatment and the requirements of authorities vary greatly from one area to the next. For repairs or maintenance, there is less objection than in the case of service mains to main sewers being laid under the paving. Once laid and tested they will seldom require any attention (although there is always the possibility of blockage in drains or in surface water gulley connections). Considerable benefits can be achieved by laying surface water and foul sewers or drains in the same trench. Some authorities employ standard 'superimposed' designs and this practice could be extended with advantage.

2 Surface elements

2.01 Introduction

Modern sewerage and drainage schemes make use of precast concrete units in many different forms, and their incorporation in the landscaped surface requires careful planning, **1**. Pipes, manholes, inspection chambers, soakaways and street gulleys are some of the products in common use. Most concrete sewage and drainage products are covered by British Standards and Codes of Practice.

The most important Standards and Codes are the following.
BS 556: Parts 1, 2: 1972 Concrete cylindrical pipes and fittings including manholes, inspection chambers, and street gulleys.
BS 4101: 1967 Concrete unreinforced tubes and fittings with ogee joints for surface water drainage.
BS CP 301: 1971 Building drainage.
BS CP 2005: 1968 Sewerage.
In terms of cover, CP 2005 recommends that sewers laid under roads or footpaths, **2**, should in general be given cover of not less than 1·2 m.

2.02 Manholes

Types

Precast concrete manholes, **3**, have now largely superseded those traditionally constructed of brickwork; they are factory made, of high quality concrete, are low in cost, quickly erected and require no maintenance.

Two types of cover are available: heavy and light duty. Heavy duty covers are for use under roads, or where similar loadings can occur. They are designed to withstand one or two wheel-loads of 112 kN at any position on the slab. Light duty covers are for use in situations other than roads (or similarly loaded locations), and are designed to withstand a single load of 35 kN in any position on the slab when supported on the outer rims.
Manholes are used to cover the access points to most services, and one form of manhole cover should be used throughout a scheme to ensure a uniform appearance.
Manhole covers and gulleys can be classified in two

1a

1b

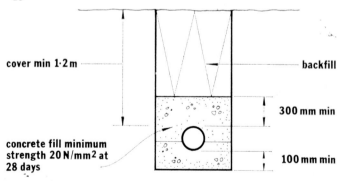

2

1 Two examples of thoughtful incorporation of surface water run-off services in hard landscaping. In addition to neat incorporation of tree surrounds in paved areas, note co-ordination of *rectangular grilles with paving pattern in middle distance of both* **a** *and* **b**. *2 Recommended detail of sewer under road or footpath.*

cover min 1·2 m — backfill
concrete fill minimum strength 20 N/mm² at 28 days
300 mm min
100 mm min

categories: the traditional cast iron **4**, patterns, and the precast concrete flush or recessed lid variety, **5**.

When detailed correctly and sympathetically, both types of manhole can offer certain benefits. Cast iron manholes and gulleys of the traditional patterns can be used to provide a certain charm and a sense of historical continuity. The recessed lid concrete variety can have the lid filled with a suitable material to match the surroundings. Only a rectangular manhole stands any chance of being aligned and positioned within a paving pattern. This is not the case with triangular, circular and 'lobed' manholes. A recessed manhole which is at an angle to the general pattern of pavings is automatically disruptive. Where a manhole cannot possibly be integrated in a paving pattern, either because of its shape or its alignment, then it should be given its own surround, **5**. With rectangular manholes, this could take the form of an edging of paving slabs laid parallel to the manhole. Another way of integrating a manhole is to surround it with a bed of in-situ concrete which follows the outline of the manhole. This could be surrounded by an area of cobbles parallel and adjacent to the paving pattern. Another solution is to replace the cobbles with plain in-situ concrete so that the square within the paving pattern containing the manhole would be one area of in-situ concrete. Manhole covers and gulleys are produced by several manufacturers in the same material and sizes as their paving. In this way, they exert minimum visual impact when designed with the paving pattern, **5**.

Design

The inclusion of a manhole or gulley often causes a structural weakness in an in-situ bay. This is due to the restraint of the free movement of the slab. Such defects show up as cracks extending from the corners of the gulleys or manholes into the slab (see Information sheet 2).

If possible, gulleys (and particularly manholes) should be placed at the end of slabs where they can be easily accommodated. If gulleys are not placed at the exact joint position, but are less than 1 m from the joint, the corner of the slab adjacent to the gulleys is likely to be weak. This weakness can be avoided by moving the position of the joint (this may involve the cutting of reinforcement and the provision of special length kerbs laid on the slab). If the gulley is only about 300 mm away from the joint, the use of a slightly larger box may be acceptable.

Manhole covers should also be boxed out (ie a rectangular concrete bay formed to surround the manhole cover), because, if settlement occurs, the trouble can then be remedied quite simply. The area of the boxed-out section should be slightly larger than the area of the manhole, as, if the slab were to rest on the manhole, any settlement of the ground around would cause cracking of the slab. When the box is filled in, a butt joint should be used, which should contain a filler to allow for horizontal movement. The joint should also be sealed to prevent the ingress of water. Neither tie bars nor dowel bars should be used between boxed out sections and the paving slab. The forms used for constructing the boxed-out sections should be designed for easy removal while the concrete is comparatively green, without risk of damage to the concrete.

2.03 Inspection chambers

Concrete inspection chambers enable drainage to be completed rapidly and with a substantial saving in site labour. They are normally installed by a pipe-laying gang and obviate the employment of bricklayers, **6**. Precast concrete inspection chambers consist of circular, rectangular or specially shaped units capable of receiving 100, 150 or 225 mm connections, **7**. Base units, **3**, with or without precast

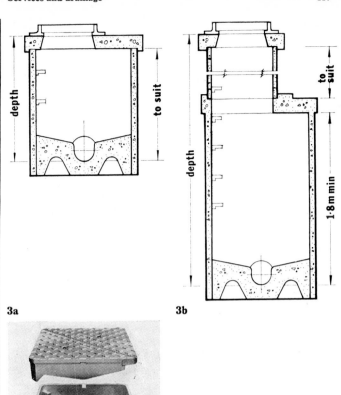

3a   3b

4

5

3 *Typical shallow and deep type precast concrete manholes. For dimensions, consult manufacturers as details vary.*
4 *Typical municipal type cast iron manhole cover and frame.*

5 *'Wexham' hydraulically pressed concrete cover and frame, developed by Cement and Concrete Association. 800 mm square frame is easily incorporated in paving layouts, and matches pressed slabs in durability and appearance.*

benching are available. The use of precast base units allows
the use of a smaller chamber than would be possible in
brick construction. Manufacturers should be consulted for
dimensions, depths and types of cover available. Inspection
chamber bases should be bedded either on 100 mm of
compacted granular material or on 75 mm of concrete—
unless the sub-soil is very stable. This can be carried out as
part of the pipe-laying operation. The chamber is then
completed by assembling the remaining components above
the base unit, the joints being set in cement mortar. With
well made cement mortar joints, the precast inspection
chamber is fully watertight, and no surround concrete is
needed. There is no need for internal rendering.
Inspection chamber covers are mainly of the light duty type.
It is recommended that the manufacturer be consulted in all
cases where vehicle traffic needs to be carried.

2.04 Street gulleys

Precast concrete street gulleys are manufactured of high
quality concrete; they can be supplied with or without traps
and provided with 100 or 150 mm outlets.
The diameters and depths available are given in BS 556,
and availability should be confirmed from manufacturers'
catalogues. The sealed type of gulley has a minimum 85 mm
water seal, rodding eyes being provided with a close fitting
concrete or galvanised stopper and chain.

2.05 Other drainage products

While all the products referred to above are described in
current standards, numerous other types of concrete
products are available to meet the drainage requirements of
engineers and architects. Some of these are described below.
Soakaways These may consist of perforated segmental
units, **8**, or tubes of various diameters with holes at regular
intervals. They have a superior storage capacity to pits
filled with rubble. Concrete cover slabs are available for all
sizes of soakaways.
Egg-shaped pipe These sewers are intended to carry a large
rate of flow when the sewer is running full. At the same
time they provide a smaller channel for the minimum flow
rate. They have the added advantage of providing greater
headroom than a circular sewer of the same capacity.
Box culverts These units are generally a rectangular section,
and are useful for carrying foul or surface water where
headroom is restricted. They have considerably larger
carrying capacity than twin or triple lines of circular pipes.
Flumes and open channels Generally semi-circular in section
and manufactured as half pipes.
Septic tanks and cesspits Often constructed of precast units.

2.06 Pavement lights

The basic purpose of pavement lights is to provide natural
daylight to basement accommodation beneath public
footpaths or other pedestrian areas. Precast concrete
pavement lights are normally situated in the public footpath
or in similar locations. In general, they are designed for a
superimposed loading to suit foot traffic. In certain
situations they may be required to carry wheel loads for
which they can be specially designed.
Pavement lights generally consist of semi-toughened translu-
cent glass lenses cast into a frame-work of reinforced concrete
ribs, **10-14**. The dimensions of the glass panels depend
upon the superimposed loading in addition to aesthetic
considerations. Glass lenses are made in a variety of types
and sizes, but for public footpaths the 100 by 100 mm size
is common and is recommended to ensure reasonable
safety against slipping. The section of the ribs and in

6

7

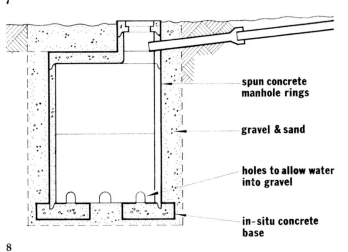

spun concrete
manhole rings

gravel & sand

holes to allow water
into gravel

in-situ concrete
base

8

6 *Precast chambers can be
installed by pipe-laying gang,
obviating employment of
bricklayers.*
7 *Typical precast inspection
chamber.*
8 *Typical precast soakaway.
For dimensions, consult
manufacturers as details vary.*

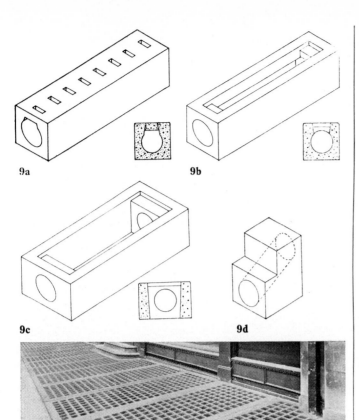

9a   9b

9c   9d

10

9 Precast drainage units for surface water removal in situations where heavy loads may be imposed on the units (eg roads and driveways). Units shown are: a standard unit; b inspection unit; c gulley unit; d outlet unit. Units a to c are 1 m long.
10 Concrete pavement lights.
11 Typical pavement light
and intermediate joint.
12 Pavement light abutting wall and alternative intermediate joint.
13 Concrete infill instead of glass lenses, for a plain surface where smoke outlets are incorporated in paving.
14 Shaped outer frame to provide keying with asphalt paving.

particular the vertical dimension varies with the superimposed loading on the panel, and with the span. The range of the vertical dimension for foot traffic loading is likely to be from 60 mm up to 150 mm. It is usual to provide a granolithic finish to the upper surface of the concrete and to incorporate carborundum in the surface to render it non-slip. Other finishes of a special character are possible and can be provided where they are considered necessary.

For most purposes, the lights are supplied with straight edges to the surround but some manufacturers shape them to suit special conditions. Where they are required, perforated metal vents can be incorporated in the lights in place of the glass lenses.

To facilitate handling on site, pavement light panels should not exceed 2·80 m². They are, in all cases, specially cast to meet the particular requirements and dimensions of each order. Standard moulds are used by each manufacturer to form the characteristic sections of his own particular

product. Manufacturers of precast pavement lights also carry out in-situ construction and they adopt this technique where it is considered to be an advantage.

Regulations governing pavement lights vary from one district to another and inquiries should be made of the local authority before design commences. In general, the projection of lights beyond the building line is regulated in accordance with the width of the pavement.

The superimposed loads which the lights must be designed to carry differ: requirements should be verified and will depend on the locality in which the lights are to be installed. A risk that the situation of the light may be subject to point load of a vehicle wheel after an accident sometimes arises and designers should allow for this possibility and design accordingly (see para 2.02, Manholes).

2.07 Smoke extracts and escape flaps

Where smoke extracts are necessary, manufacturers can provide panels to meet such regulations. The indicating plates necessary to mark their position can, if required, be cast into the concrete.

Single leaf escape flaps, which are hinged and counter-balanced to meet the prescribed requirements, can be supplied to comply with most regulations.

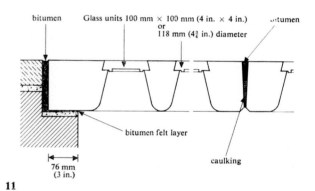

11

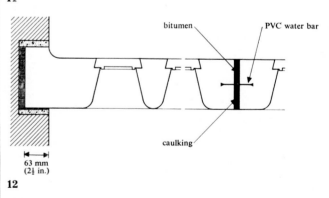

12

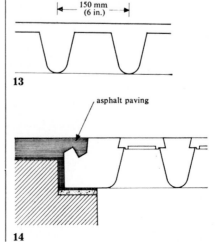

13

14

Information sheet 7
Material specification

1 Specification methods

1.01 Traditional versus performance specification

With traditional methods of specification the writer accepts responsibility for interpreting function and performance in material and production and the contractor who has to follow the specification cannot be held responsible if the result is more expensive than it would have been if alternative materials and techniques had been used. Nor is the contractor liable if the product does not perform as the specification writer intended.

A performance specification, on the other hand, specifies neither the material to be used, nor the method of production, but sets out precisely the standard required, allowing the contractor to select materials and production techniques. When used as a basis for competitive tendering a real money incentive is provided for the contractor to develop economic methods of production to satisfy the standard required.

1.02 Samples

To overcome the deficiencies of traditional methods of specification a useful technique is to specify a concrete finish to be as already achieved elsewhere. Alternatively a prototype can be produced as a basis for tendering, although the appearance of concrete in a small sample may be vastly different from that in larger areas. At all events the main purpose is to provide each contractor with an accurate guide for tendering purposes, **1a,b**.

The terms of the contract should oblige the contractor to produce trial samples. The samples should be reasonably large for in-situ work and include at least one construction joint so that they can form the basis for an agreed 'minimum' standard of quality for the whole contract. Such trials will also serve to familiarise operatives with the techniques necessary to produce an acceptable finish. If this procedure is adhered to, detailed information on how to achieve the required finish need not be made part of the contract; it will be provided as informal advice for the contractor who can then determine his own methods in the light of experience.

2 Materials

2.01 Cement

All cements should comply with the requirements of the relevant standards; see table I.

The colour of normal grey cement depends upon the colour of the raw materials from which it is made. To ensure uniformity of colour, cement should be obtained from one works, and if particular uniformity of colour is of importance, from one day's production. In the latter case the supplier should be advised well in advance to ensure that adequate facilities are available to deliver the cement in one consignment.

2.02 Coloured finishes

Powder pigments or paste pigments are available for addition to concrete or mortar with ordinary Portland, white Portland, or other Portland cements. The powders are finer than ordinary Portland cement. Up to 10 per cent pigment is added, depending upon density, covering capacity and the tint required. With white Portland cement, clearer tints are produced on addition of light coloured pigments. Pigmented concrete will normally exhibit the same properties as unpigmented concrete. There may be a small loss of compression strength at early stages when significant additions of some pigments, such as carbon black, are made to concrete.

2.03 Aggregates

In general aggregates should comply with the requirements of one of the following British Standards:
BS 882; 1201; BS 877; BS 1047; BS 3797; BS 4619.
The designer may specify or approve on request the use of other aggregates, including types or gradings not covered by one of the above British Standards, provided there are satisfactory data on the properties of concrete made with them.
When high strength concrete is required the source as well as the type of aggregate may need careful selection based on the results of trial mixes.
Where it is known that any property of any aggregate is likely to have an unusual effect on the strength, density, shrinkage, moisture movement, thermal properties, creep, modulus of elasticity or durability of concrete made with it, the engineer should take account of these factors in the structural design and in the workmanship requirements.
Aggregates having a high drying shrinkage, such as some dolerites and whinstones, and gravels containing these rocks, produce concrete having a higher drying shrinkage than that normally expected. This can result in deterioration of exposed concrete and excessive deflections of reinforced concrete unless special measures are taken. For further information see BRS Digest 35.
Special precautions should be taken to ensure that aggregates are free from impurities, such as pyrites, which may cause staining. A useful guide is to make test samples using the proposed materials. The source of supply should not be changed.
Uniformity in colour of the coarse aggregate is important only for exposed aggregate finishes. The grading and particle shape of the aggregate should be kept constant otherwise necessary adjustments in mix proportions for workability will result in colour changes. The preferred nominal maximum sizes of coarse aggregate are 40 mm, 20 mm, 14 mm and 10 mm.
Where variations occur in the shape or colour of the aggregate from the same source, arrangements may have to be made to stockpile a sufficient quantity of selected aggregate for the contract.
A major factor in the delivery times for precast concrete with special finishes is the availability of the facing aggregate. The difference in delivery times from the various quarries is substantial, and can vary from weeks to months from the order date. Trade associations and research bodies have extensive knowledge of local conditions and should be consulted at the outset.

Table I Standards for various cement types

Type	In accordance with
Ordinary Portland cement	
Rapid-hardening Portland cement	
White Portland cement	BS 12
Coloured Portland cement	
Portland blastfurnace cement	BS 146
Low heat Portland cement	BS 1370
Sulphate-resisting Portland cement	BS 4027
Low heat Portland blastfurnace cement	BS 4246
High alumina cement	BS 915
Supersulphated cement	BS 4248
Ultra high early strength Portland cement	The requirements for the physical
Water repellent Portland cement	properties for ordinary Portland
Hydrophobic Portland cement	cement given in BS 12

Where cements other than those complying with the requirements of BS 12 or BS 146 are used, account should be taken of their properties and any particular conditions of use.
Neither high alumina nor supersulphated cements should be mixed with any other type of cement.

1a

1b

1a *Sample panel showing different effects achieved on same panel by four indirect finishes: sand blasting, top left; chiselling, top right; bush hammering, bottom left; and point tooling.*

1b *Sample panel showing different colour effects achieved by using different cements: on left, white Portland cement, and on right, ordinary cement.*

Aggregate types

The characteristics of all major aggregate sources are now known and both gravel and crushed rock aggregates are suitable for pavement surfaces. There is some evidence that crushed rock aggregates produce higher flexural strengths than gravels. The mechanical properties of the aggregate are of prime importance; particularly toughness, which may be defined as the resistance of aggregate to failure on impact and is related to the crushing value. Table IV gives average test values for British stones of different groups. In all cases low is best.

Aggregates should not contain deleterious materials in such a form or in sufficient quantity as to affect adversely the strength or durability (including resistance to frost) of the concrete. For example soft coal is undesirable and, if finely divided, may disturb the process of hardening. At one end of the scale, anthracite, which is chemically inert, does not normally reduce concrete durability although it powders on exposure and could produce a surface blemish. Brown lignite and fossilised wood readily disperse in the concrete mix and can cause local stains and weak zones. Limits have been prescribed as shown in Table II.

Sand won from the sea-shore and river estuary contains salts, and opinions differ as to its acceptability. Certainly special care is required with deposits just above the high water mark where a high concentration of salt is sometimes found.

Soft chalk particles may reduce the frost resistance of concrete at the surface and an arbitrary limit is therefore applied in special circumstances to the calcium carbonate content, although this term embraces hard chalk and limestone which are generally physically and chemically sound. It is reasonable to impose a maximum limit of 1 per cent by weight on mica content when considerations of strength or durability indicate need for control.

Research in the last ten years has drawn attention to drying shrinkage and Table III gives the average results of tests on groups of crushed rock aggregates and on concrete.

Information on aggregates

British Ready Mixed Concrete Association, Shepperton House, Green Lane, Shepperton, Middlesex TW17 8DN (Shepperton 43232).

British Quarrying and Slag Federation, 14 Waterloo Place, London, SW1Y 4ER (01-930 7107).

Cement and Concrete Association, 52 Grosvenor Gardens, London SW1 (01-235 6661) and Wexham Springs, Slough, Bucks (Fulmer 2727).

Department of the Environment, Building Research Establishment, Garston, Watford, Herts (Garston 74040).

Department of the Environment, Building Research Establishment, Scottish Laboratory, Kelvin Road, East Kilbride, Glasgow (East Kilbride 33941).

Department of the Environment, Road Research Laboratory, Crowthorne, Berks (Crowthorne 3765).

Departments of geography and/or geology in local universities and technical colleges.

Institute of Geological Sciences, Exhibition Road, South Kensington, London SW7 (01-589 9441).

Institute of Quarrying, 62-64 Baker Street, London W1 (01-486 2547).

Sand and Gravel Association of Great Britain, 48 Park Street, London, W1Y 4HE (01-499 8967).

3 Mix design
3.01 Principles

Mix design is the selection and proportioning of the various materials used to produce concrete so that the desired properties are achieved in both plastic and hardened state.

Table II Maximum coal and clay content of aggregate

Type of particle	Maximum content (per cent by weight)	
	Fine aggregate	Coarse aggregate
Coal	1·0	1·0
Clay lumps	1·0	0·25

Table III Average test values for various aggregates and for concretes made with the aggregates

	Water absorption of aggregate (per cent)	Drying shrinkage of concrete (per cent)	Moisture expansion concrete
Limestone	0·5	0·03	0·02
Granite	0·3	0·03	0·02
Dolerite	1·4	0·05	0·04
Basalt	1·4	0·06	0·05
Flint	1·0	0·03	0·02
Quartzite	0·8	0·03	0·02

Table IV Average test values for various aggregate stones

	Crushing value	Impact value	Attrition
Basalt	12	16	4
Flint	17	17	3
Granite	20	13	3
Limestone	24	12	6
Quartzite	16	16	3

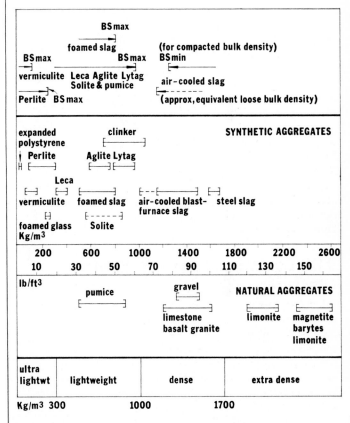

3 *Loose bulk densities of coarse aggregates. Densities laid down by 1974 BSs are shown in top row.*

As the properties of many of the materials used in concrete are inherently variable, in order to achieve a satisfactory mix the proportions of materials should be verified by trial mixes on the site. At present mix design is based on considerations of strength, durability and workability and frequently the prime concern is to satisfy these basic requirements at minimum cost. However where appearance and uniform colour are desired it is recommended that the mix be richer.

In general, if surface blemishes are to be avoided, the aggregate should consist of a fine sand with a continuously-graded coarse aggregate of the largest practicable maximum size. The concrete should have the maximum feasible workability and the mix should be designed to have the maximum compressive strength and the lowest economic water/cement ratio.

In the past, specifications for concrete prescribed proportionally nominal volumes of cement, fine aggregate and coarse aggregate,'5. Owing to the inherent variability of the materials used, concrete produced with a fixed aggregate/cement ratio and a given workability varied widely in strength. For this reason a minimum compressive strength was included in many specifications. However, this type of specification makes mix design too inflexible and concrete unnecessarily expensive where good materials are available, and adequate strength impossible to achieve when unsuitable materials are used. This type of specification provided neither opportunity nor incentive to design a mix better suited to the materials available and the needs of particular jobs. The tendency nowadays is for specifications to be less restrictive, laying down limiting values.

For concrete designed to have a special property or to produce a particular surface finish it is important to specify the requirements in detail and where possible to state the reasons for any special requirements so that the contractor may more fully appreciate the object of the work. In appropriate circumstances any of the following information may be included but great care should be taken to ensure that the requirements specified do not conflict with each other:

Design mix for special structural concrete
Grade designation
Minimum cement content
Nominal maximum size of aggregate
Required workability
Maximum cement content
Required brand or type of cement
Required source or special type of aggregate
Required admixture
Air content of fresh concrete
Maximum or minimum temperature of fresh concrete
Rate of sampling and testing
Other requirements

3.02 Continuous and gap-graded concrete

When concrete is made with aggregates which contain proportions of particles of all sizes from the largest to the smallest, the result, **4b**, is an overall grading which is continuous. When concrete is made with aggregate in which particles of an intermediate size are almost completely missing, the result is an overall grading which is gapped, **4a**. When gap-grading is used intentionally, say for an exposed aggregate finish, the coarse aggregate usually consists of one single sized material. The main advantage on site from using gap grading is that only two sizes of aggregate have to be handled compared with three or more for continuous gradings. Gap grading may also lead to a more economical mix if the available supplies of aggregate tend naturally to provide that type of grading. However, workability and cohesiveness in gap graded concrete are far more sensitive to small changes in water/cement and coarse/fine aggregate ratio and particular care must be taken to ensure that the concrete is fully compacted and does not segregate. It has been found that there is no significant difference in the

Table V Grades of concrete

Grade	Characteristic strength	Lowest grade for compliance with appropriate use
	N/mm²	
7	7·0	plain concrete
10	10·0	
15	15·0	reinforced concrete with lightweight aggregate
20	20·0	reinforced concrete with dense aggregate
25	25·0	
30	30·0	concrete with post-tensioned tendons
40	40·0	concrete with pre-tensioned tendons
50	50·0	
60	60·0	

The characteristic strength is that determined from test cubes at 28 days for concrete with any type of cement excluding high alumina cement concrete. For high alumina cement see Section **12**.

4a

4b

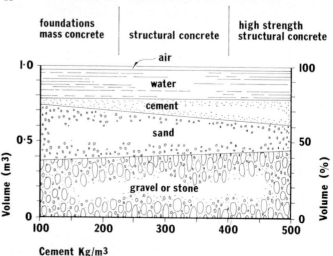

4a *Gap-graded concrete with mix consisting of 50 mm to 20 mm pebbles, with aggregate exposed by washing and brushing three to four hours after casting, then cleaned down with hydrochloric acid.*

4b *Continuous graded concrete, with aggregate exposed by shot-blasting.*
5 *Range of concrete mixtures and applications. Graph assumes ordinary Portland cement, 20 mm gravel and 50 mm slump.*

compressive strengths of concrete mixes of a given richness and workability whether continuously graded or gap graded aggregates are used, provided that the optimum proportion of fines is used in each case.

In terms of appearance, whether a continuous grading or gap grading is used depends upon the finish required. For example where a bold exposed aggregate finish is desired a gap-graded aggregate is preferable because where a continuous grading is used the large aggregate tends to become distributed in an uneven manner over the face of the concrete, resulting in a less uniform finish. Conversely, for the production of 'fine' exposed aggregate finishes the best results are usually obtained using continuous graded aggregates up to 10 mm size.

4 Admixtures

4.01 Uses

With good materials, correct mix design and thorough compaction there should rarely be any need to use admixtures. Admixtures cannot correct imperfections in mixing, handling or placing concrete, but they may sometimes be used with advantage to improve workability, increase durability, inhibit efflorescence and provide colour retention, reduce risk of frost damage or aid the curing of concrete in low temperatures.

In any event, admixtures should be employed only for specific purposes, such as:
● accelerating the rate of hardening;
● increasing the workability;
● retarding the setting processes;
● entraining air;
● producing aerated concrete;
● decreasing the permeability of hardened concrete;
● reducing the rate of heat evolution;
● colouring concrete.

4.02 Types

A material other than the usual concrete ingredients can be introduced into a mix by employing:
● commercially available chemicals, industrial by-products or the like;
● proprietary admixtures;
● special cements in which the material is incorporated as an additive.

Before acceptance, an admixture should be fully evaluated by preliminary tests which are appropriate for the circumstances. For example, it may be necessary to determine its effect on the strength or workability of a particular concrete if one or other of these is a significant property. In most cases, comparative tests with and without the admixture or with two different admixtures of the same type will be valuable, although care is required in interpreting the results.

When a proprietary material is used, the supplier should be expected to furnish the following information:
1 type, ie main effect;
2 name of manufacturer;
3 name of product or trade name;
4 description, eg is product in liquid form?;
5 main active ingredients;
6 manufacturer's typical range of dosage in grammes or litres/100 kg cement;
7 percentage anhydrous $CaCl_2$ by weight of cement at maximum dosage;
8 percentage additional air entrainment at maximum dosage;
9 does segregation occur in storage?;
10 is product affected by freezing?;
11 'shelf life' of product when stored in accordance with manufacturer's recommendations;
12 are precautions necessary when using marine aggregates containing chlorides?;
13 compatibility with cements;
14 possible effect of over-dosage.

Air entraining agents improve the resistance of concrete to frost action, increase durability and the cohesiveness of plastic concrete. Air can be entrained by adding appropriate agents at the mixer. Only approved entraining agents should be used, otherwise variations in strength and air content are difficult to regulate.

To obtain optimum frost resistance without excessive loss of strength the concrete should contain between 3 and 6 per cent of air by volume. When using air-entrained concrete it is necessary to modify the mix proportions to allow for some loss of strength.

Accelerators increase the rate of hardening of cement and are used mainly to improve the rate of gain of strength at low temperatures. Calcium chloride has been well-proved in practice. It should not be used in pre-tensioned, pre-stressed concrete, nor in concrete subject to steam curing, nor be mixed with sulphate resisting or aluminous cement.

The quantity of calcium chloride added to any mix must be carefully controlled; normally quantities vary between 0·5 and 1·5 per cent of the cement weight. When used to excess calcium chloride will impair the durability of concrete.

Other admixtures consist of many proprietary materials designed as plasticisers, retarders and waterproofing and gas-forming agents. The use of these materials must be very carefully considered not only in terms of mix design but also with regard to possible complications on site arising from the false sense of security given to operatives who may imagine that such admixtures can make good any defects in workmanship. The effect of these types of admixture will depend on many variable factors and the only real way to assess their value is by trial mixes.

Some minerals such as chert are capable of reacting with the alkalis in cement but in this country these are fortunately not present in concreting aggregates.

Information sheet 8
Appearance and weathering

1 Appearance

The appearance of concrete depends mainly on three factors: colour, texture and pattern.

1.01 Colour

Both the *cement* and the *aggregate* used will contribute to the colour of the concrete unit; but their relative importance will depend to a large extent upon the treatment given to the concrete face.

Cement
While natural grey cement can be combined satisfactorily with various aggregates, use of white or coloured cement greatly extends the range of possible colour combinations.

Aggregate
In choosing an aggregate for exposed work, durability and availability must be considered as well as colour. The natural gravels and crushed rocks normally used as concrete aggregates are generally durable, easily available and provide a range of colours to suit most purposes (colours vary considerably according to geological type, and even within each type). Limestones and slates can provide a variety of colours, depending upon the district of the quarries. Crushed brick, tile and other manufactured materials such as glass can be used for special effects but since they require crushing and grading their cost is generally higher than that of natural concrete aggregates.

1.02 Texture

The appearance of concrete owes as much to texture as to colour, and these two properties must be fully considered to obtain the most effective results. The variety of textures available in precast concrete is very wide, ranging from a polished surface to that obtained by exposing 70 mm to 100 mm size uncrushed aggregate such as cobbles.
There are four important factors to be considered:
● surface area to be covered;
● distances at which the surface will normally be viewed;
● aspect (orientation);
● particle shape and size of the aggregate; and reflectivity of aggregate.
The larger the surface to be covered, the greater the distance at which it will normally be viewed, and the larger and/or the more reflective the aggregate, the bolder will the texture probably be. Conversely, a small surface viewed from nearby will probably have a relatively fine-textured surface, which will in turn necessitate a small-sized aggregate.

1.03 Pattern

Pattern can be obtained in a variety of ways and will be affected by treatment of joints, shape, arrangement and use of 'profiled' areas. The usual method of producing a pattern is by treatment of joints, which can be made very fine or emphasised on the surface.

The plasticity of concrete as a material enables 'profiles' (with deep relief or geometrical patterns) to provide interesting play of light and shade, and adds greatly to the range of visual effects that can be produced. Really bold effects can be obtained by the use of such materials and also by the juxtaposition of units of different size, shape, texture and colour. Infinite variations can be built up in this way.

2 Weathering

Regardless of the type of finish to be used, it is essential to consider at the outset the effects of weathering. The scope of this important design process can be defined under four headings:
- internal stability of constituents;
- surface nature;
- environment;
- design and detailing.

2.01 Internal stability of constituents

Cement: the white deposit or lime bloom so commonly seen on concrete surfaces is caused by the movement of calcium hydroxide, liberated by hydration of the cement, and carried in solution to the surface of the concrete where carbonation and evaporation take place. Incidence of lime bloom is largely regulated by the permeability and texture of the concrete surface and weather conditions.

Aggregates: probably the most serious form of internal damage is contamination by the iron pyrites often found in certain gravel aggregates. Although the stability of the concrete may not be seriously affected by their presence, discoloration (dark brown) can be considerable.

Sand from the sea shore or from a river estuary contains salt and special care is required when deposits are removed from just above low water mark, as any surplus salt will absorb moisture from the air and cause efflorescence. The simplest course is to wash the sand in fresh water.

Steel: provided that the reinforcement is fixed so that the cover is in accordance with the relevant standards and codes and the concrete is fully compacted, staining caused by rusting of reinforcement should not occur. However, staining can also be caused by wire ties being too close to the surface.

2.02 Surface nature

Permeability: the permeability and absorptive capacity of concrete affect its life and weathering, and for surface finishes these characteristics are as important as strength. Obviously the damage will be determined largely by the degree of penetration permitted by the nature of the concrete; the more impervious the surface the less deterioration there will be.

Unfortunately, complete impermeability is impossible as concrete is a porous material. Voids are formed in any of the following three ways.

1 Excess water in the mix. For complete hydration of cement a water/cement ratio of 0·22 to 0·25 is required. However, in order to achieve enough workability for complete compaction, the water/cement ratio must be considerably more than this, so there is always uncombined water present in the mix. This excess water eventually leaves passages for water or fluid to penetrate.

2 Incomplete curing (allowing the concrete to dry out prematurely as hydration proceeds) will increase the porosity because set cement paste occupies less volume than fresh paste, regardless of its water/cement ratio.

3 Air is trapped during mixing and placing and a small proportion of it can never be removed, even by very prolonged vibration. Thus it is apparent that for concrete with maximum durability and low permeability the water/cement ratio should be reduced to a minimum consistent with enough workability for full compaction. Special attention to curing can further reduce the permeability of concrete. Admixtures are sometimes used to make concrete less permeable but, in general, good mix design followed by full compaction and thorough curing are the best guarantee of high resistance to penetration. (Some admixtures can provide colour retention which may be difficult to achieve merely by good mix design.)

Smoothness: the surface of concrete as cast in a mould consists of a dense hard layer of fines and cement paste. This surface is less absorbent than the majority of traditional building materials and is apt to craze and to weather unevenly, particularly if the mix is too rich in cement or has too high a water content. Mould faces with a matt finish result in less crazing and less discoloration. Where a smooth or polished concrete finish is required it is best produced by grinding and polishing after the mould is struck, although it must be remembered that this is a very expensive technique.

When rain runs down a smooth profiled concrete face it flows across any recessed areas on the surface producing light-coloured streaks. Water discharging over any element should be foreseen at the design stage and channelled so that the resultant staining highlights the visual effect, accentuating any modelling of the surface.

Roughness: perhaps the least exploited property of concrete is its plastic nature when in a fluid state. As a general rule the more textured and profiled the concrete surface, the better it will appear and the better it will weather, as minor blemishes and defects can be minimised by skilful design. With the advent of plastics, textured and contoured surfaces can now be produced at a price which is highly competitive with that of traditional materials. A great advantage of these new techniques is the improved weathering and low maintenance of strongly profiled and moulded surfaces. With rough-textured and profiled surfaces the rain running across the face is broken up and distributed over the surface, so that streaking is minimised. Another advantage of the rough-textured units is that they can absorb a certain amount of moisture. In the case of walls built of dense and relatively non-absorbent materials, rain falling on the surface streaks down the outer face; there may actually be greater moisture penetration than in the case of a rough-textured unit and more water may concentrate at the joints than with a more absorbent unit, creating the possibility of discoloration at these points. Because an exposed aggregate surface provides ample room for water to expand, damage to the surface is less likely if the water freezes. Consideration must however be given to the amount of bedding of the different shapes and sizes of stone in order to ensure that they remain in place after years of weathering. It is recommended that the stones are embedded in the matrix to a depth of at least two-thirds of their volume.

2.03 Environment

Chemicals: in practice serious chemical attack on concrete is rare. Resistance to the forms of attack most often encountered (leaching-out of cement, sulphate action and the presence of slightly acidic water in the atmosphere) generally depends more on the properties of the concrete mix than on the quality of the cement. In extreme cases however the advice of cement manufacturers should be sought.

Frost: perhaps the most significant hazard to durability is the action of frost on both fresh and hardened concrete. Frost has two distinct effects on fresh concrete. First, if the water within the mix freezes there is an overall increase in volume, causing disruption. Second, the frozen water is not available for hydration of the cement so that setting and hardening of the concrete are delayed. To avoid this problem the specification should include a clause to the effect that the temperature of the concrete should not be allowed to fall below 4°C during placing and curing, and until the concrete has attained a strength of at least 5 N/mm². The effect of frost action on hardened concrete comes from penetration of water into exposed surfaces, and it normally manifests itself as scaling. When the temperature of saturated concrete is lowered, the absorbed water begins to freeze and exert pressure in two basic forms: increase of volume (approximately 9 per cent) and hydraulic pressure (dependent on the resistance to flow encountered by the excess water). The effect of this type of damage can be reduced to negligible proportions by using a relatively high cement content at a low water/cement ratio so that as much water as possible is used to combine with the cement during hydration, and sufficient strength is developed to resist stresses set up during freezing conditions. In addition the concrete should have low permeability and low absorption to minimise penetration of water and fluids. Where extreme environmental circumstances are encountered the use of air entrainment or waterproofing admixtures will greatly improve frost resistance.

Abrasion and wear: the resistance of concrete to abrasion and wear is not normally a major consideration. However, as a general guide, resistance to abrasion is directly related to strength, so that the higher the crushing strength the greater the resistance to abrasion. When the compressive strength of concrete exceeds 42 N/mm² the type of aggregate has little effect on abrasion resistance.

Pollution: generally this is a problem in urban and heavy industrial areas. In such situations light-coloured finishes will rapidly become grimy, and there will be a marked contrast between the appearance of sheltered surfaces (where dirt collects) and the appearance of well-washed surfaces exposed to prevailing wind and rain. This contrast can add to the visual interest if very skilfully exploited, and designers should give careful attention to likely weathering patterns. Aggregates of angular shape and crystalline surface texture will generally hold more dirt than aggregates of rounded particle shape and glossy texture. Concrete surfaces can be cleaned by scrubbing and spraying with water in exactly the same way as natural stone buildings.

3 Removal of stains

A number of procedures and products have been developed for removing or reducing the intensity of many stains on concrete; but it is far preferable to avoid such staining in the first place. Mud and dust can be removed with a stiff brush and water, but stains resulting from efflorescence cannot be removed at all, and the various chemical treatments described below are all difficult processes which may lead to disappointing results.

In all cases, remedial work should not be attempted until the concrete is at least three or four weeks old, as the stains often disappear during this time without further treatment.

3.01 Cleaning difficulties

Difficulties have three causes: concrete surfaces are generally permeable so that both stains and applied cleaning solutions tend to become absorbed in the surface layers; concrete is chemically reactive and dissolves in some cleaning solutions more readily than the chemicals causing the stain (for example, rust is more chemically inert than concrete); and finally, surfaces are often left as cast, so that any locally applied treatment may alter the texture or colour sufficiently, to cause a blemish. When chemical means are neither appropriate nor effective, consideration can be given to mechanical treatment (eg grinding, polishing, grit-blasting, hacking out; see table I) or to covering the stain with a suitable coating.

3.02 Chemical methods

The materials used for removing stains from concrete include acids, organic solvents, emulsifying agents and bleaches as well as other substances outside these classifications. To combat the tendency of the concrete to absorb liquids the materials are sometimes mixed into a paste or poultice with talc, whiting, kieselguhr (diatomite) or other powder. Applied to concrete, the mixture draws the stain out to the surface. This technique is not always appropriate with acids and the powder must not react with the acid. Before treatment is attempted as much as possible of the material causing the stain should be mopped up or scraped off the concrete surface.

Corrosive and poisonous chemicals must be handled with care. It is advisable to wear rubber gloves, goggles and other protective clothing during application.

Acids: these are often used to remove a thin layer of laitance or mortar from a surface which is supporting a stain such as brown iron oxide (rust). With light etching a superficial stain can often be removed without any marked effect on the texture of the concrete. In addition to hardened cement paste, limestone and marble aggregates are dissolved by acids. This fact is exploited in some terrazzo floor cleaners which contain a small amount of an organic acid to improve the efficacy of the products.

A cheap, powerful, but very corrosive chemical cleaner is hydrochloric acid. In the concentrated form it is sold as a 36 per cent solution of the gas in water with a specific gravity of 1·16. This solution is usually diluted 20 times with water to give '5 per cent hydrochloric acid solution' before it is applied to concrete. Common names are spirit of salt and muriatic acid.

Other acids used for cleaning are formic acid, citric acid crystals dissolved in water, acetic acid, phosphotic acid and oxalic acid crystals dissolved in water (which is a very poisonous solution). Some of these form the bases of proprietary cleaners and stain removers. A typical formulation might contain 15 per cent citric acid crystals dissolved in water containing a wetting agent.

Organic solvents: these may include petrol, benzene, xylene, carbon tetrachloride and trichlorethylene. Mixed with powdered talc or the like these can be used to remove bitumen and oil stains. Some solvents are highly flammable: others give off vapours which can be toxic over a long period.

Table I A comparison of mechanical methods

Method	Relative speed	Relative cost	Advantages	Disadvantages
Water spray	Slow	Low	No risk of damage to masonry except under frost conditions. No danger to public or operatives. Quiet.	Limestone may develop brown, patchy stains. Water penetration may damage interior finishes, hidden timber and ferrous metals. Some risk of drain blockage. Possible nuisance from spray and saturation of surrounding ground. Often requires supplementing with an abrasive method or high-pressure water lance.
Dry grit-blasting	Fast	High	No water to cause staining or internal damage. Can be used in any season.	Risk of damage to surface being cleaned and to adjacent surfaces, including glass. Cannot be used on soft stone. Possible noise and dust nuisance. Risk of drain blockage. Injurious dust from siliceous materials. For best results, needs to be followed by vigorous water washing.
Wet grit-blasting	Fast	High	Less water than with water spray method. Less visible dust than with dry-grit blasting.	Similar to dry grit-blasting but greater risk of drain blockage. Some risk of staining limestone. Can result in mottled finish if operatives are unskilled.
Mechanical cleaning	Fast	High	No water to cause staining or internal damage. Can be used in any season.	Considerable risk of damage to surface, especially mouldings. Injurious dust from siliceous materials. Hand rubbing may be necessary for acceptable finish.
Hydrofluoric acid preparations	Medium	Low	Will not damage unglazed masonry or painted surfaces. Quiet.	Needs extreme care in handling—can cause serious skin burns, and instant damage to unprotected glazing and polished surfaces. Scaffold pole ends need to be plugged and boards carefully rinsed.
Caustic alkalis	Fast	Low	Rapid cleaning of some types of limestone with minimum use of water.	Needs extreme care in use; can cause serious skin burns and damage to glazing, aluminium, galvanised surfaces and paint. Only preparations covered by an agrément Certificate should be used and only in accordance with the terms of the Certificate. Incorrect use can cause serious progressive damage to masonry.
Steam cleaning	Slow	Medium	No risk of damage to masonry except under frost conditions.	As 'water spray' but with less risk of staining. Not easy to obtain uniformly clean appearance.

Emulsifying agents: many surface-active agents serve as emulsifying agents for oils and greases and this property can be of value in the development of techniques for cleaning contaminated concrete because the emulsified material can be washed away with water. Several efficient proprietary emulsifying liquids have been developed for the motor trade as engine cleaners and are safe to use on concrete.

Detergents, including washing-up liquids, have this emulsifying property as well.

Bleaches: the two bleaching materials which are generally available are bleaching powder and household bleach solution. These are essentially calcium hypochlorite (milk of lime) and sodium hypochlorite solution. These solutions bleach out most beverage and ink stains and will kill green algal and fungal growths which disfigure external concrete surfaces.

Other materials: salts such as sodium citrate, sodium metasilicate, dichlorophen and products with unidentified active constituents may also be employed for treating stains on concrete.

Rust marks can be removed by acid but it should be remembered that this treatment will affect the surface of the concrete itself. It is not advisable to grind or wire brush the surface as permanent marks on the concrete can result. Considerable care should always be exercised during these operations and no attempt should be made to spread the area of treatment in order to 'fade out' the effect of staining. The best method is to test a sample of the proposed removing agent on an inconspicuous area. Its value can then be assessed and the composition and strength varied accordingly. This technique is particularly important when 'fillers' are used to form a paste. The condition of the surface will influence the choice of method and it is important to remember that accumulated weathering on old surfaces disappears when stains are removed.

Generally speaking, stain prevention is better than cure. All removal methods described must be regarded as inferior to prior precautions that would obviate the need for them.

3.03 Mechanical methods

For mechanical methods, see table 1 above

Information sheet 9
Surface finish: brush and wash exposed aggregate

1 Introduction
1.01 General
Several advantages may be obtained by removing the outer skin of cement and fine material which normally forms on the surface of concrete, thus exposing the aggregate.

An exposed aggregate finish revealing the true nature of the material and colour can be provided in the final finish by careful choice of aggregate and cement. Exposed aggregate finishes also tend to weather in a more attractive manner than comparable finishes where the hardened skin of cement paste is retained.

The main disadvantage of this finish, regardless of casting method used, is lack of uniformity in colour and texture not only from panel to panel but also within a single panel.

2 Brush and wash technique
The cheapest and simplest method of exposing the aggregate for both in-situ and precast concrete is to brush and wash, or spray the concrete while it is still sufficiently soft, **1a,b**.

1a

1b

1 *Two simple techniques for removing outer skin of cement paste, and exposing* *aggregate:* **1a** *brush and wash;* **1b** *compressed air and water jet.*

Treatment demands the early removal of formwork as soon as the concrete is mature and strong enough to be self-supporting.

2.01 Exposure technique

The usual time for treatment is between two and six hours after casting. During normal weather conditions (temperatures between 12°C and 18°C), using ordinary Portland cement, the latest time at which effective treatment can be carried out will usually be between 16 to 18 hours after casting.

Exposure should begin round the perimeter of the slab to the required depth before the remaining surface is brushed. The depth of exposure obtained in any particular case will depend on the grading of the aggregate and the workability of the concrete.

Either stiff bristle or wire brushes should be used, with plenty of water, to remove the matrix and clean each piece of exposed aggregate. Brushes should be constantly washed in clean water to prevent any clogging which would reduce their effectiveness.

After the whole area has been evenly exposed, the surface should be sprayed with water and lightly brushed down, cleaning away all adhering mortar and revealing the true colour of the aggregate.

Any cement bloom left on the exposed aggregate finish can be removed with a weak solution of 10 per cent commercial hydrochloric acid. This treatment should be delayed for at least three (preferably seven) days, and the cleaned surface must be washed down afterwards to remove any traces of acid.

2.02 Supervision

Successful aggregate exposure depends upon first class formwork, a high degree of mix control and correct placing and compaction of the concrete. If the concrete mix varies from batch to batch, or if mortar is allowed to leak out, the appearance of the concrete after the aggregate has been exposed will be patchy. The removal of the cement skin from the face of concrete reveals any defects in both the formwork and the concrete, and accentuates any deficiencies.

Skill is required during the brushing operation, and to ensure uniform depth only the minimum of material should be removed to achieve the required effect. It must be emphasised to all operatives that even exposure of the aggregate is essential, and that the temptation of brushing too vigorously in one area must be avoided, for it may then be impossible to obtain the same depth of exposure over the whole face in the time available.

2 Moulds were coated with retarding agent, and demoulded panels are now being brushed and hosed to expose aggregate.

If retarders are used, **2**, application should be closely scrutinised and be in strict accordance with the manufacturers' recommendations. Contamination of the body of the concrete is always a danger and extreme care must be exercised at all stages of the work.

Concreting should be continuous for each section, otherwise irregular daywork lines will show on the face of the exposed aggregate finish. Concreting cannot be stopped at a random position without a stop-end to be continued later, even though the interval extends merely over a lunch break. All such construction joints and stop-ends must be carefully designed and detailed.

The sequence of the work must be carefully planned as it is essential to brush the surface of the concrete as soon as possible. The surface mortar normally begins to harden when exposed to the air and only a limited time is available during which the aggregate can be exposed.

2.03 Curing

Under certain conditions there is danger of thermal shock to the concrete when cold water is used for washing the surface. As soon as the final operation is complete, all exposed concrete surfaces must be cured with an appropriate system.

3 Materials
3.01 Aggregates

Finished appearance depends largely on type, size and shape of aggregates, and on colour of constituent materials.

It is most important when choosing aggregates for various colour effects to ensure that they are also durable. Crushed rock and gravels are normally used and have the advantages of durability, availability and a range of colours suitable for most purposes. Consider the degree of exposure to which the different shapes and sizes of stone are subjected to ensure that they remain in place after years of weathering. It is recommended that the stones are embedded in the matrix to a depth of at least two-thirds of their mean size.

3.02 Coloured cements

To merely tone down or darken the surface of the concrete, incorporate 2 to 4 per cent of ferric oxide by weight of cement in the mix during the mixing process. Proprietary materials are available which are specially prepared for this purpose. A trial area should be laid, as the effect obtained will vary with the particular material used.

For coloured work, coloured cement can be obtained from the manufacturers and the use of a coloured aggregate to match the cement will assist in providing a uniform appearance. It is desirable to lay a small trial area in order to ensure that the colour is as desired.

When constructing coloured concrete surfaces, the general principles of concrete work apply to the construction of the slabs. It is usual, however, to adopt two-layer construction, using the more expensive coloured cement only in the top 50 mm of the slab; separate tools and mixers should be used for each course. About 15 per cent extra cement should be used in the mix to compensate for the amount of pigment in the coloured cement. When using coloured cements it is specially important to keep the proportions of the mix constant, as variations will cause difference of tint in the surface.

Good curing is essential when using coloured cement. Waterproof paper or plastics film should be used as they will not stain the concrete.

Information sheet 10
Surface finish: tooled

1 Introduction

1.01 General

One of the most common methods of increasing the texture of in-situ concrete finishes is to tool the surface, **1**. Basically the several different ways of tooling concrete all have the effect of removing the surface skin of hardened cement paste and exposing the aggregate in the body of the concrete below.

For normal exposed aggregate finishes tooling has been superseded largely by abrasive blasting techniques (see next information sheet). However where deeper textures or special aggregate effects are required tooling may still be an appropriate method. Many different textures can be achieved by working the surface with a variety of tools.

1.02 Appearance

Removal of the skin of hardened cement paste from the face of the concrete reveals, rather than masks any imperfections, whether they be from poor formwork design or lack of attention to mixing, placing and compaction of the concrete. This cannot be emphasised too strongly as the most common misconception is that tooling will improve the appearance of poor quality concrete.

2 Tooling techniques

The final finished appearance of the concrete will depend upon the detailed design, the composition of the concrete, and the method of tooling used to expose the aggregate. There are three main categories of tooling: point tooling, bush hammering, and chiselling.

1 *Tooling process being carried out by Kango hammer.*

2.01 Point tooling

With this method the concrete is pitted overall with a tool known as a 'point', which is either power or hand operated. When done by hand it is known as 'picking'.

The size and distance apart of the pits can be varied according to the effect required, but they must be sufficiently close together for the whole of the skin of hardened cement paste to be removed, eliminating any smooth patches of mortar between pits. The size of the pits will depend on the amount of pressure, the length of time the point is held in one place and the type of point tool used. Point tools can have either a short drawn point where a light texture is required or a long drawn point to achieve a deep texture.

2.02 Bush hammering

This form of tooling is most commonly used in Britain. It gives the concrete a rough texture, varying according to the type of coarse aggregate used and whether the surface is lightly or heavily hammered. The cement skin is crushed or spalled off with a toothed hammer head, generally power-operated. A roller bush hammer tool (a development of the bush hammer holder and disc) is available for use with a mechanical hammer and is faster in action than the bush hammer disc. This type of bush hammer has now largely displaced the original disc-type head. The normal thickness of material removed from the face of the concrete using either of these heads is about 3 mm, although by going over the surface more than once a greater thickness of material can be easily removed.

2.03 Chiselling

Several finishes can be grouped under the heading of chiselling, including those made with such tools as the boaster, the broad tool, drafting chisel, and claw chisel.

3 Precautions

The following precautions should be observed for a successful finish.

3.01 Design and detailing

The first requirement for all exposed aggregate finishes is a good quality, well compacted concrete in watertight formwork. The final exposed finish will be only as good as the cast concrete.

Formwork must be carefully made and watertight. If joints are poor and mortar leaks through, the joint will show up in the finished work. Since construction joints will show in any case, their location should be carefully planned before construction begins, wherever possible concealing them by combining them with a design feature.

The treatment of arrises requires careful detailing and although it may be possible in certain circumstances to carry hand tooling right up to an arris, some spalling will inevitably take place which is extremely difficult to remedy. Methods of avoiding this problem include:
● leaving a margin at all arrises of at least 20 mm (preferably 40 mm) by drawing a line or masking with battens;
● forming round or chamfered edges;
● designing the cast face which is to be tooled as a projection, leaving the margin as a recess. This can be done by fixing a 6 mm timber fillet at the edges of the formwork to form a margin up to which the tooling can be carried out.

3.02 Specification*

It is possible to tool concrete made with most aggregates although some are more suitable than others for particular techniques.

Natural aggregates are inclined to shatter under the blows of a hammer. This can lead to bond failure between the aggregate and the matrix, so that some of the particles become loose and fall out, leaving unsightly pits in the face of the concrete.

The best aggregates for tooling are those which can be cut or bruised without fracturing. Most of the igneous rock and hard limestone aggregates are eminently suitable.

The method of surface tooling and the colour tone required can determine the type of aggregate used. If a *chiselled* finish is required, the marks of the tool are an important feature of the design and the aggregate would normally be a softer type such as limestone. With *bush hammering* the marks of the tool are relatively unimportant, the colour and texture of the finished surface being the deciding factors. In general when concrete is tooled the bruising of the aggregate considerably lightens its colour.

3.03 Reinforcement cover

For all exposed aggregate finishes the cover to reinforcement, as laid down in the relevant BS codes of practice, should be increased to compensate for the amount of material removed from the face of the concrete. The usual recommendation is that cover is calculated from the back face of the largest size of aggregate used in the mix.

3.04 Supervision

In general, to produce a satisfactory tooled finish it is necessary first to produce a good plain finish following the main recommendations given for exposed concrete. It is important to obtain complete compaction and to avoid leakage through joints because these effects are always noticeable. To achieve the best finish, concreting must be continuous for each section; otherwise irregular daywork lines will show on the face of the exposed aggregate finish. Concreting cannot be halted at a random position without a stop-end, and then continued later, even though the interval merely extends over a lunch break. All such construction joints and stop-ends must be carefully designed and detailed.

For the most consistent results care must be taken to ensure that:
● small areas are completed at a time;
● tooling is not carried out in lines across the surface;
● uniform pressure is applied when power tools are used;
● minimum and uniform depth of exposure is achieved.

Power tooling is recommended when there is a large area to be tooled and an appreciable thickness of concrete has to be removed. Hand tooling is best for relatively small areas and should always be used for areas close to an arris.

Skill is required during the tooling operation to ensure uniform depth of exposure and only the minimum amount of material should be removed to achieve the desired effect. Even exposure is essential. If this point is not clearly made the tendency will be to tool vigorously in one spot and then find that it is impossible to obtain the same depth of exposure over the whole face.

It is recommended that work begins on the area placed first and the depth of exposure obtained in this section will determine the degree of exposure over the remainder of the surface.

*For advice on choice of aggregate and use of coloured cement, and on curing, see Information sheets 2 and 7.

Information sheet 11
Surface finish: abrasive blasted

1 Introduction

1.01 General

Compared with other methods of exposing aggregate, abrasive blasting offers many advantages, and it is therefore becoming increasingly popular.

● Unlike brush and wash, acid etch, or the use of retarders to expose the aggregate, all of which must be carried out at an early stage while concrete is still soft; and unlike tooling, which cannot be carried out for at least three to four weeks after casting, abrasive blasting can be done over a comparatively wide time range. The production programme is therefore less complicated, and the exposure process probably more economical.

● Unlike tooling, abrasive blasting can be used right up to and around arrises without danger of spalling.

● Uniformity of colour and texture is more easily obtained with this technique than with any other; while at the same time decorative effects can be produced by varying depth of exposure, or by leaving some areas untreated in regular or irregular patterns, **1**.

The main disadvantage of the process is that the exposed skin is bruised by the abrasives, resulting in a matt finish.

1 *Patterns of various depths being carved by blasting.*

1.02 Appearance

To achieve uniform exposure, use single size, gap graded concrete with aggregate of the largest practicable maximum size. This tends to overcome the problem, associated with all exposed aggregate finishes, of non-uniform distribution of particle size throughout the cast concrete. The depth of exposure of the aggregate can be varied by the amount, nature and type of grit used, and by the length of time the blast jet is retained on the surface. The texture of a sand blasted finish is different from that of a grit blasted finish, and again, from a shot blasted aggregate. Shot blasting is not normally recommended as particles of shot may lodge in the exposed surface and cause staining. It is important that the type, nature and origin of abrasives are not changed, as such variations will affect the uniformity of the exposed finish.

The surface aggregate exposed by abrasive blasting appears weathered and slightly bruised yet retains the fresh and sparkling appearance of natural rock colouring. There is a slight change to a greyer appearance, however, when black or dark glossy aggregates are grit blasted.

2 Technique
2.01 Blasting

Abrasive blasting techniques are normally carried out by specialist firms and it is essential that the type of finish and in particular the depth of exposure be clearly specified. The action of the blast jet first removes the laitance covering the surface aggregate and then erodes the softer matrix from between the aggregate particles. The depth of exposure should not exceed one-third of the mean size of the aggregate, otherwise aggregate may become displaced. The stage at which abrasive blasting should be carried out depends largely on the type and size of aggregate, the cement used, and on the time of year at which the concrete is placed. The important thing is to commence blasting as soon as possible provided the concrete has achieved the requisite maturity and strength. Normally on concrete made with ordinary Portland cement, abrasive blasting should be commenced between 16 hours and three days of casting.

To obtain a consistent finish the whole of a job should, if practicable, be treated at the same age.

The blast jet of grit and compressed air is directed at the surface to be treated with the discharge nozzle placed about 300 mm away from the work surface. On plain areas with concrete less than three days old the rate of exposure can be at the rate of 4 m^2 to 6 m^2 per hour. With age hardened concrete the effective coverage drops to about one-half this rate.

It is quite usual for the whole of the exposed surface to be given a second sand blasting at the end of the job to remove any dust and mortar droppings, finally blowing the surface clean with a jet of air. In some instances the whole of the exposed surface is then given a silicone treatment to make the concrete water-repellent.

2.02 Design and detailing

The first requirement for all exposed aggregate finishes is to produce a good quality, well compacted concrete in watertight formwork. Construction joints unless carefully considered will show, so it is important that details are decided upon before construction begins. Wherever possible the joints should be concealed by combining them with a design feature.

In contrast to tooled finishes, the treatment of arrises does not require such careful detailing, as it is possible to carry abrasive blasting right up to the arris without spalling.

As with all exposed aggregate finishes, the cover to reinforcement as laid down in the relevant codes of practice should be increased to compensate for the amount of material removed from the face of the concrete. As a guide, the minimum cover before treatment for any exposed aggregate finish should be at least 45 mm.

2.03 Specification

It is important when choosing aggregates for various colour effects that they are also durable. Crushed rock and gravels are normally used and have the advantages of durability, ready availability and enough colours for most purposes. Consider the degree of exposure given to the different shapes and sizes of stone to ensure that they remain in place after years of weathering. It is recommended that exposure of the surface does not exceed one-third the mean size of the aggregate.

Remember that the colour of the aggregate darkens when exposed to natural weathering cycles.

2.04 Supervision

Uniformity of the finished surface depends to a very great extent upon the degree of supervision exercised at all stages of the job, as a high standard of workmanship is essential to achieve an acceptable finish.

In general, it is necessary first to produce a good plain finish following the main recommendations given for exposed concrete. It is most important to obtain complete compaction and to avoid leakage through joints because these effects are noticeable to a considerable depth. If joints are poor and mortar does leak through, the joint will show up in the finished work.

To achieve the best finish it is necessary for concreting to be continuous for each section, otherwise irregular daywork lines will show on the face of the exposed aggregate finish. Concreting cannot be halted at random without stop-ends, and then continued later on, even after a lunch-break interval. All such construction joints and stop-ends must be carefully designed and detailed.

The scale of work must be carefully planned as it is essential to expose the surface of the concrete as soon as possible after casting. The surface mortar normally begins to harden when exposed to the air and only a limited time is available during which the aggregate can be exposed without excessive expenditure of blasting material.

Even exposure is essential. If this point is not clearly made, the tendency will be to linger in one spot and then find that it is most expensive to obtain the same depth of exposure over the whole face.

Operations should commence on the area placed first and the depth of exposure obtained in this section will determine the degree of exposure over the remainder of the surface. Extreme care must be exercised when sand blasting the junction between concretes of different ages. At such points a strip 30 mm to 40 mm wide is left until the less mature concrete has reached the required hardness before it is blasted. By this means it is possible to make construction joints virtually invisible.

2 *New improved helmet offering greater safety and comfort for the grit blaster.*

Information sheet 12
Surface finish: plain

1 Initial finishing

When the surface of the slab is to be the wearing surface, the initial finishing is an important first step in the production of a hard-wearing and level surface. The object is to bring the compacted concrete to within specified level tolerances, and with a closed smooth surface without excessive working, so that the final finishing operation, concentrates on the finish and not on correcting defects in level or surface.

All ridges and steps in the surface should be eliminated, and any irregularities should be of long wavelength, ie a curvature of not more than 2 to 3 mm in 600 mm. This is particularly important when a grinder is to be used, so that the machine can easily follow the surface profile. It is simply achieved by reasonable attention to surcharges during compaction and by careful use of the compacting beam, **1,** followed by a regulating pass with a skipfloat. When a power-trowelled finish is required, the initial finish can be achieved by first power-floating as an alternative to the use of the skipfloat. A small amount of hand work along joint

1

1 *The double beam screed vibrator enables one man to strike-off, vibrate, compact and level concrete in a single pass.*

2 *The scraping straight-edge.*

2

edges is also necessary to ensure full compaction and elimination of lips and spillage of mortar on to the adjacent hardened concrete.

1.1 Hand floating and brushing

Where a high-grade surface is not required and traffic is light, the initial finishing with the skipfloat may be adequate, or a brushed texture may be applied with a bristle broom to give improved slip resistance. Alternatively the surface may be finished with a hand wood float which will leave a slightly coarse texture.

2 Finishing techniques

The prerequisite for good performance from a concrete surface is full compaction, and this is usually fulfilled by means of a single- or double-beam compactor which is fitted with a vibration unit. The action of surface vibration tends to draw water to the surface, and wet finishing may produce a weak layer at the top. If this is not removed, the hardened surface will have low durability and wear resistance, and a high risk of dusting. The following techniques may be used with most of the construction methods which are included later.

2.1 Hand trowelling

Traditionally slabs are finished by successive trowelling with steel trowels, with a delay of an hour or more between each trowelling to allow further moisture to evaporate. Surplus weak mortar should be scraped from the surface as trowelling proceeds, and by exerting considerable physical pressure on the trowels during subsequent trowellings to close the pores in the surface left by the evaporated moisture, a skilled man can produce an excellent finish. Less skilled trowellers tend to trowel in the mortar and to use less pressure, and a poor wearing surface may result. The dusting of neat cement on to the surface, or wetting down before trowelling is also bad practice which can produce surfaces which may scale, dust or craze, and have generally low wear resistance.

2.2 Power floating

A power float may be used as a preliminary to power trowelling to regulate and close the surface, **3**. Floating must not be started until surface moisture has evaporated and the concrete is stiff enough to take the weight of the machine, or else a weak surface and poor levels will result.
Power floats are also used for some techniques involving broadcast finishes (see Information sheet 13).

2.3 Power trowelling

No mortar is removed during power trowelling and timing is therefore critical if a hard-wearing surface is to be achieved. Surface moisture must evaporate, and the concrete stiffen sufficiently to take the weight of machine and operator, **4**. This frequently requires concreting to stop around midday so that finishing can be completed during working hours, otherwise long periods of overtime may become necessary.
After further moisture has evaporated, successive trowellings with the blades tilted at greater angles produce a hard-wearing surface. Power trowelling may also be used to prepare the surface for the direct application of thin sheet coverings.

3

4

5

6

3 *The initial floating process produces a flat, even, but coarse textured surface on the concrete.*

4 *The final process is to carry out power trowelling which produces a dense, close, flat, smooth and very hard wearing surface of excellent quality and appearance.*

5 *Vacuum dewatering unit, self contained or compressor powered, uses sieve cloths and airtight top cover.*

6 *Grinder at work finishing a warehouse floor. Grinding is done 3 to 4 days after the concrete has been placed.*

2.4 Vacuum dewatering

The timing problems of power trowelling can largely be overcome by the use of the vacuum dewatering process immediately after the initial compaction of the slab, or topping. The surface is covered by a rigid or flexible suction mat which incorporates filter layers and which is connected to a vacuum generator, **5**. An 80 to 90 per cent vacuum is applied at the rate of two to three minutes per 25 mm of depth, and a considerable quantity of water is withdrawn from the concrete (the filters prevent the loss of cement). The effect is to cause the concrete to stiffen rapidly and within an hour or so of being placed the surface may be finished by power trowelling. The process also improves the wear resistance and strength of the concrete.

3 Early-age power grinding

The power grinder may be used to produce either a hard-wearing surface, or a surface suitable for the direct application of thin sheet coverings. Grinding is a finishing technique, and is not intended to correct gross irregularities in the surface. The initial finishing is important so that the machine may work uniformly over the surface to remove the thin weak surface layer, say about 1 mm thick, and expose the basic strong concrete, **6**. Sudden steps and lips should therefore be avoided, and irregularities should be of long wavelength. Grinding should be carried out dry and while the concrete is young. For maximum economy this should be within the period from two to seven days after laying, when grinding rates of 40 m²/h, or more, should be possible. It should begin as soon as the concrete is sufficiently hard for sand particles not to be torn from the surface. The precise timing depends upon mix proportions and ambient conditions, but timing is far less critical than for power trowelling, and overtime working is unnecessary. Grinding at later stages becomes progressively slower and less economic. The technique should produce an overall 'glass paper' texture without excessive exposure of coarse aggregate, and give good wear and slip resistance.
It should be noted that the concrete grinder is a relatively slow-speed machine, unlike the high-speed terrazzo grinder.

Information sheet 13
Surface finish:
broadcast (scatter)

1 Introduction

The usual method of achieving an exposed aggregate finish is to spray and brush the surface some hours after casting (see Information sheet 9). The surface of a structural slab may be upgraded for more severe duties by the introduction into the surface of selected natural, or metallic, aggregates usually blended with cement, **1**. The pre-blended material is broadcast over the surface at a suitable rate immediately after the initial compaction of the slab, and is then compacted in with a power float and finally finished by trowelling. This is a specialised technique which should only be entrusted to those firms which have the necessary skills.

1 *Selected dust-free aggregates scattered on to concrete surface, prior to tamping in and exposing by water spray.*

Table 1 Recommended thickness and bay areas for high-strength concrete toppings

Type of construction	Thickness of topping (mm)	Bay areas	Comments
Monolithic	12 to 20 (allow 3 to 4 mm surcharge)	As for structural slab	Topping laid within 3 h of finishing slab
Bonded	20 to 40 (allow 4 to 8 mm surcharge)	Maximum width 4·5 m Maximum area 25 m² Length/ breadth 1·5	Preparation of slab as described in I..S.2 May be laid in long strips, and sawn into bays of 25 m² area.
Unbonded	Not less than 75 mm	Not more than 10 m²	Some curling is likely even with limited areas. The use of 20 mm maximum size coarse aggregate may help to limit curling. To reduce it further, over-slabbing 100 to 125 mm thick may be used.

2 Cement-based toppings

Where the abrasive conditions require concrete of very high strength, and it becomes uneconomic to use this grade throughout the full depth of the slab, or because of site difficulties (ie the incorporation of service duct cover frames into the surface), a high-strength topping may be used. Similarly a sand-cement topping may be required for decorative and service reasons, when site conditions make it impossible to direct-finish the slab. In all cases, the topping should be used in conjunction with the appropriate structural grade of concrete. Toppings are constructed in one of three ways (see Table 1 above).

2.1 Monolithic construction

This may be used for high-strength. The smallest thickness of topping may be used, between 12 mm and a maximum of 20 mm and, in this form of construction, the top course of the structural concrete is compacted with a notched beam to leave the surface below finished level by this amount, **3**. The topping is then placed and compacted, preferably with a double vibrating beam, within not more than three hours of finishing the structural concrete. The topping becomes part of the structural thickness of the slab and this form of construction is the soundest and most economical of the alternatives involving toppings. The construction method, bay sizes and arrangement of joints follow those for the slab itself.

2.2 Bonded construction

If monolithic construction cannot be adopted, the topping must be bonded to the slab after it has hardened. In this case, the topping must not be included in the structural depth of the slab.

Toppings, being thin, have little structural strength, and their satisfactory performance depends upon their being fully bonded to the slab. It must be emphasised that cement-based toppings (and most resin-based ones) have high shrinkage movements which will lead to debonding, curling and cracking if an adequate mechanical key is not provided as described below.

Preparation of the structural slab
The weak surface laitance of the slab, and any other contaminating material, must be removed completely and the coarse aggregate exposed cleanly before the topping is applied, by using mechanised plant such as the pneumatic scabbler. This operation is facilitated if the slab is left

smooth and not ridged with tamp marks. The additional use of bonding agents, or admixtures, is of little purpose unless this preparation is carefully and completely done.

Topping construction
To maintain the required thickness of topping, screed rails should be fixed to the slab to line and level. They should be timber battens, or steel angles, well bedded or bolted down (especially if vibrating-beam/compactors are used) and

2

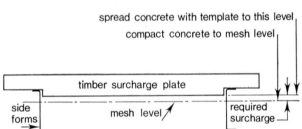

spread concrete with template to this level
compact concrete to mesh level
timber surcharge plate
side forms
mesh level
required surcharge

3

4

2 Aggregate is tamped and trowelled.
3 Notched timber template used to produce an even concrete surcharge before
compaction to mesh level.
4 Dry mortar for felt floating is trowelled on to fresh backing concrete.

should be aligned where possible with joints in the slab. All dust and debris should be removed with an air jet or vacuum, and the surface thoroughly wetted. Immediately before the topping is placed, surplus water should be removed, and a cement slurry of the consistency of cream brushed into the surface. The topping should be placed before the slurry dries out.

In general, particular attention should be paid to ensuring full compaction, especially at edges and corners. All joints in the slab should be continued through the topping, and a maximum bay width of 4·5 m is recommended for ease of laying and accuracy of levels.

An alternative technique is felt floating. A 10 mm thick bed of fairly dry mortar, with selected aggregates included, is trowelled on to the freshly compacted concrete, and a felt pad is then drawn across the surface in successive operations to pick up the cement and finer particles from the surface, 2. This gives a final appearance of a much finer texture than the ordinary scatter technique; but so much skill is involved that the felt pad method tends to be uneconomic.

2.3 Unbonded construction

Toppings are unbonded when a damp-proof, or isolating membrane, or an insulating layer is interposed between the topping and the slab. When an old surface which is heavily contaminated with oil or grease requires re-surfacing, debonded construction may be necessary, and an isolating membrane of plastics sheeting, or building paper, should be laid on the slab before the topping is constructed. The sheets forming the membrane should be lapped 50 mm at all joints.

Movement joints in the slab should be carried through the topping whenever possible to avoid the risk of reflected cracking.

Unbonded construction is the least satisfactory form of construction and, despite increased topping thickness and severe limitation of bay areas, some curling must be anticipated. This is often unacceptable, and a better solution would be to overslab with structural concrete finished with a monolithic high-strength topping, together having a thickness of 100 mm or more.

5 *After felt process is complete the topping should be trowelled to ensure a level finish.*

Information sheet 14
Surface finish: profiled

1 Introduction

For in-situ work, interesting finishes can be produced by treating the upper surface of the concrete after full compaction, before setting takes place. A variety of textures can be produced by rolling, tamping, sawing, or brushing the surface of the fresh concrete. It is, however, difficult to avoid slight differences in colour and texture using these techniques, so they should not be employed unless such variations in appearance are acceptable (or possibly even desirable).

Uniformity of colour and texture depend on uniformity of constituent materials in the mix; uniformity of mix proportions; uniformity of temperature and humidity during curing; and uniformity of time at which the technique is applied, and of manner of application.

1.01 Rolled finishes

The various rolling techniques all produce a rippled finish, caused by the fine materials being drawn by the roller into ridges on the slab surface, **1**. If the pattern is sufficiently

1 *A facing slab finished by rolling with a 50 mm tube.*

bold, and the joints between adjoining units accentuated by suitable detailing, techniques of this kind have the advantage of minimising visually the effect of colour and textural variations in the concrete.

On wide panels the largest practicable diameter of roller should be used, a tube of about 150 mm and stiff enough to avoid deflection. Such rollers will also help in levelling the concrete to the required tolerances and regularity— provided the roller is worked off the side members of the mould, and provided the correct quantity of concrete has been fully compacted in the mould (any excess will tend to flow under the roller and leave a wave behind it).

Two types of roller are available for imposing a texture on the surface of plastic concrete.

The first is a light roller having twin drums of expanded metal 2, each with a diameter of 127 mm and a width of 1·1 m. The second is a steel pipe of 300 mm diameter, 3·5 m long and weighing 44·5 kg per metre, around which sheets of patterned rubber or expanded metal can be wrapped.

The first of these is generally thought to produce a less satisfactory surface than the second type, which can be fitted with various patterns of rubber sheeting, the most satisfactory being a ribbed pattern producing a deep texture with an average depth of groove of 5 or 6 mm. Or it can be fitted with one of two types of expanded metal: general purpose diamond mesh with 30 mm by 12 mm apertures; or similar, but flattened, mesh. The flattened mesh is considered less satisfactory than the ordinary; and it produces a shallower and less uniform texture.

When rolled over a concrete surface, a mesh-covered cylinder of the recommended type produces a texture which is emphasised if repeated passes are made. The texture is characterised by protruding 'fins' formed through the apertures in the steel, and generally thought to be satisfactory in appearance, but there are two defects. First, the fins tend to consist of weak mortar, and therefore will probably be fairly quickly worn down by traffic. Second, a ridge of mortar forms at the side of the roller, which becomes more pronounced with repeated passes.

1.02 Tamped finishes

With this technique the concrete is compacted by a vibrating beam of the type used for semi-manual road construction (see Information sheet 2), and the final finishing pass is carried out without vibration to give texture to the concrete.

A ribbed surface can be produced by tamping with the edge of a board or beam, 3. Alternatively, a tamper with grooves formed along the length of the working edge (rather like a comb) can be pressed into the surface following an edge template fixed to the side forms.

This technique requires a mix with high sand content to achieve a profiled surface without voids. With some types of profiled screed boards, which are dragged across the surface, 4, it is necessary to place a layer of mortar on the fresh concrete to produce a reasonable profiled finish without tearing the surface. The amount of surcharge required is critical.

2 Twin-drum articulated mesh cylinder.
3 Ribbed surface being produced by tamping with the edge of a board.
4 Profiled tamping beam being worked off side forms, to produce heavily textured surface.

5 Dragged finish, produced by drawing a profiled board across the concrete surface.

2

3

4

5

1.03 Sawn finishes

Another variation to the tamping technique is to work a straight-edge across the surface of the fresh concrete with a sawing action. The main disadvantage is that the finished panels will tend to vary in pattern. To a certain extent this can be controlled by working the straight-edge between profiled side form members which govern the spacing and depth of grooves, and keep the lines parallel so that the pattern will conform to that of adjacent units on the building.

1.04 Other finishes

Other textures may be achieved by brushing or by scoring the surface of the concrete, before it has hardened, with such tools as a steel trowel or length of reinforcement to give a regular or irregular pattern.

The texture depth achieved by brushing the surface with a soft or medium-soft bristled broom generally does not exceed 0·5 mm, **6**; and although such textures are satisfactory when new, their life under heavy traffic is short.

Broom types

● Soft broom with 450 mm wide head, and 150 mm long brass bristles. Two degrees of texturing can be achieved, by applying either normal or heavy pressure.
● Wire broom with 450 mm wide head, and spring steel tape bristles 0·3 mm thick, 1·25 mm wide, about 100 mm long, and arranged in two or four rows of tufts.
Probably the most satisfactory broom type is that with two rows of spring steel tapes; and this is specified by the Ministry of Transport specification for road and bridge works, **7**.

Tined units

These can be used to achieve a deeper and hence more durable rough texture than that produced by brushing. The individual tines can be designed to form suitably shaped parallel grooves and ridges in the surface of the concrete; and it is essential that they induce aggregate into the ridges so formed, rather than surface laitence only.
A commercially available lawn rake comprising curved rubber tines, **8**, is suitable for such finishes. A minimum pitch of 12 mm is desirable; 15 mm is preferable; and 18 mm is probably excessive.

1.05 Vibration

The application of vibration to the texturing devices described above is necessary to achieve satisfactory penetration into the surfaces of less workable concretes.

Where methods of texturing the surface are not carried out when the concrete is in its plastic state, **9,** then the alternative technique is to employ mechanical machine groovers.
There have been significant developments in the design of groove cutters and the latest machines offer a variety of methods which range from reflex hammer motors to tungsten carbide-tipped flails or tines, **10, 11**. Cutting speeds vary from 8-10 metres a minute for limestone based aggregate to 4-5 metres a minute for flint aggregate concrete.
Quite often the use of such machines offers the only economic alternative to grooving the concrete in its plastic state.

6 *Brushing fresh concrete surface by means of bristle broom;* **7,** *the same operation* *carried out by wire broom; and* **8,** *by rubber-tined rake for very heavy texturing.*

6

7

8

9

10

11

12

9 *Cement & Concrete Association's prototype machine for texturing concrete during laying.*
10 *After concrete has hardened, grooves up to 10 mm deep and 3 mm wide may be cut at 30 mm centres.*
11 *Self-propelled machine which cuts grooves 6-8 mm wide and 4-6 mm deep.*
12 *Scabbling device for an overall exposed aggregate finish on hardened concrete.*

Information sheet 15
Surface finish:
ground and polished

1 Introduction

1.01 Cost

Because of the cost and time taken by mechanical grinding and polishing of concrete surfaces, this technique is not common in the UK except in precast work of the terrazzo type. It is not, of course, practicable to grind large areas of in-situ concrete.

The cost of polishing and grinding procedures is dependent mainly upon the hardness of the aggregate and the depth of exposure required.

1.02 Methods

There are two methods: dry and wet grinding. Wet grinding is preferable because the ground-off material works up into a paste, and this helps the grinding action. When dry grinding is used, operatives must wear masks to protect their eyes and lungs from dust.

1.03 Appearance

Initially, successful grinding and polishing depends upon first class formwork, a high degree of mix control, and correct placing and compaction of the concrete within the forms. If the concrete mix varies from batch to batch, or if mortar is allowed to leak at corners or between boards, the appearance of the concrete after treatment will be patchy. The removal of the cement skin from the face of concrete reveals any defects in both formwork and concrete, often accentuating deficiencies.

Uniformity of the finished surface depends to a very great extent upon the degree of supervision exercised at all stages of the job. It cannot be over emphasised that a high standard of workmanship is essential if an acceptable finish is to be achieved.

1.04 Weathering

Perhaps the most obvious blemish associated with smooth impervious concrete surfaces is that any resultant low spots on the finished concrete will be accentuated by water movement across the face. No matter how well grinding and polishing are carried out, the final weathering will depend upon the nature and durability of the cast concrete. Unfortunately polished finishes tend to craze and crack and are not normally recommended for external conditions of weathering.

2 Design

2.01 Design and detailing

Normally concrete which is to be given a ground or polished finish should be cast against forms suitably lined to produce a smooth surface free from any defects (eg steel, grp, film-faced plywood, or plastic-faced ply as was used for St Catherine's College, Oxford).
Formwork must be carefully made and watertight. If joints are poor and mortar leaks through, the joint will show up in the finished work.

2.02 Specification

The type of aggregate will have a distinct bearing on the result where a polished finish is required, so it is most important to select an aggregate that can be ground economically and will take a polish. The most suitable aggregates for this type of finish are marble, spar and some classes of limestone. True granites will take a high degree of polish but the labour involved makes the grinding of granite aggregates very expensive.
Almost all natural gravels are unsuitable for polishing, but they can be used where only a ground finish is required.

3 Site procedures

3.01 Supervision

The first grinding using a coarse stone removes the skin of rich cement mortar from the face of the concrete, exposing the aggregate and any air pockets hidden below the surface. Plenty of water should be used during the grinding process and the operative should have a bucket of clean water and a soft brush so that the cement paste worked up by the grinding process can be cleaned away and the depth of exposure ascertained. When the skin of mortar has been removed from the face of the concrete to the required depth, the concrete should be washed down to remove all laitance from the surface. Twenty-four hours after the first grinding, all air holes and other blemishes in the surface should be stopped and filled with a fine mortar consisting of cement and a similar sand to that used in the body of the concrete, but having been passed through a 600 microns sieve. The proportion of cement to sand used should be similar to that used in the concrete, for example $1:1\frac{1}{2}$, if a $1:1\frac{1}{2}:3$ mix was used. When stopping and filling ordinary Portland cement concrete it is advisable to blend the cement used for this purpose with white cement, as any stopping and filling, unless toned down with white cement, will show up. When the filling is thoroughly hard, usually seven days later, the surface should again be wet ground, this time using a medium carborundum stone. The final grinding and polishing can be achieved with a fine carborundum stone. If a ground or semi-polished finish only is required then it is unnecessary to carry out the third grinding and polishing operation, it being a matter of choice whether the second grinding is done with a medium carborundum or fine stone.

3.02 Curing

In normal weather conditions the first grinding should be done some 24-36 hours after stripping and special care must be taken to prevent any damage to the arrises.

1

2
1 *Metal sprayed on to cast concrete units and subsequently polished to provide a metallic concrete finish.*
2 *Finishing the concrete surface of a motorway bridge by means of a carborundum wheel.*

Information sheet 16
Surface finishes:
Aggregate transfer and
sand-bed methods

1 Introduction

1.01 Aggregate transfer method

The process of aggregate transfer consists of sticking selected aggregate to the rough side of sheets of pegboard, positioning the prepared sheets within the formwork and then filling the shuttering with concrete. On removal of the formwork, the pegboard is stripped from the face of the concrete and the aggregate exposed by brushing.
Water-soluble cellulose compounds have been found to give satisfactory results in sticking the aggregate to the pegboard. When mixed with water and a sand filler they produce a consistency like thick cream. To increase the bond strength of the adhesive mixture, up to 15 per cent of hardwall plaster may be added to the filler.
The prepared mixture should be evenly spread on the pegboard to a depth of one-third the mean size of the aggregate, for example 13 mm for 40 mm aggregate.
The thickness of the glue/sand should be carefully controlled by the use of a notched float. Aggregate should be sprinkled on the glue/sand bed so as to cover the surface completely, and should be tamped into the mixture by means of a wood float with sponge rubber. When this operation has been completed the liner should be tilted to allow surplus stone to fall away and, if necessary, any irregular places should be made good. When 25-40 mm or larger aggregates are used, the individual stones should be placed by hand, and each stone pressed into the glue/sand mixture. The prepared panels must be allowed to dry and harden under warm conditions for at least 36 hours before use.
This finish is not recommended for in-situ work and is not widely used in precasting as the sand-bed method (see 1.02) is cheaper and provides an equally good finish.

1.02 Sand-bed method

For precast work the sand-bed technique is often used with face-down casting to obtain an exposed aggregate finish, particularly when using aggregates larger than approximately 40 mm in diameter. The method is to place the aggregate by hand in a layer of dry sand spread evenly over the base of the mould. The thickness of the sand-bed determines the degree of exposure of the aggregates and the deeper the sand the more exposure there will be.
To ensure a proper bond it is essential to cover the backs of the stones with a workable cement and sand mortar, followed immediately by the normal backing concrete.
Skilled labour is required. Hand placing of large aggregate is usually slow (a rate of about 2m² per man hour), and further delays are caused during concrete placing because

normal vibration cannot be used (it could cause the stones to become displaced and the mortar to run through to the face of the slab).

Choice of the sand for the bed is most important as it will provide the background colour of the matrix and preferably should be the same colour as the sand used in the backing concrete mix.

1.03 Where to use

Perhaps the most significant advantage of the aggregate transfer method is that if the aggregate has been correctly applied to the pegboard liner the final exposed finish will conceal minor deficiencies caused by formwork, variations in mix proportions, or inadequate compaction.

2 Appearance

2.01 Weathering

It is most important when choosing aggregates for various colour effects to ensure that they are also durable. Crushed rock and gravels are normally used and have the advantage of durability, good availability and a range of colours suitable for most purposes.

3 Specification and curing

3.01 Specification

Adhesives which have been found satisfactory are the water-soluble cellulose compounds commercially known as 'Cellofas B extra high viscosity granular', 'Polycell' and 'Adwata'.

It is essential to ensure that any adhesive will not permanently stain or discolour the final exposed aggregate finish. The choice of sand used as filler will affect the background colour and the final overall appearance of the surface. The sand may be derived from the same materials as the aggregate to be exposed, or it may be of a different nature and colour depending upon the effect required.

3.02 Curing

Liners are usually stripped the day following the placing of the concrete but in cool weather it may be necessary to allow more time.

After the removal of formwork and stripping of liners, the adhesive and sand covering the face of the aggregate can be removed by scrubbing with a bristle brush and then washing down the surface with clean water.

Any cement bloom left on the exposed aggregate finish can be removed with a weak solution of 10 per cent commercial hydrochloric acid. This treatment should be delayed for between three and seven days and the cleaned surface washed down afterwards to remove any traces of acid.

1

2

3

1 *Aggregate transfer method: placing aggregate on glue/sand bed which has been laid on pegboard to a carefully controlled depth, usually one third mean diameter of aggregate.*
2 *Pegboard liner, with glued-on aggregate, is now inserted into mould, with temporary baffle in front to* *protect aggregate during placing of concrete.*
3 *Alternative to aggregate transfer: Sand-bed casting. Selected aggregate has been placed on sand-bed in mould, and layer of cement/sand mortar is now being placed over aggregate. Mortar will ensure bond between aggregate facing and concrete backing.*

Information sheet 17
Metals and concrete

1 Metal types

1.01 Steel

If bare steel is immersed in an alkaline solution it will acquire a thin oxide film which protects the metal against further corrosion; and provided the solution remains alkaline, the steel will not corrode.

Concrete provides such an alkaline environment, and while the aggregate type, cement content, water/cement ratio, degree of compaction and method of curing all play an important part in determining the degree to which the steel will corrode, the compressive strength of the concrete is a useful indicator: the higher the strength of the concrete, the lower will be its permeability, and the less the danger of corrosion.

1.02 Zinc-coated steel

Zinc-coated reinforcement has been employed successfully to reduce the risk of corrosion. Tests have shown that the initial attack on zinc by the alkalis released during the hydration of the cement is not progressive, and that the coating can be expected to have good durability. After the concrete has hardened, attack usually ceases, provided that the concrete is kept dry.*

1.03 Stainless steel

Stainless steel does not denote a single alloy, but a group of alloys consisting of something like 60 varieties, so that the term stainless steel should not be taken too literally. Stainless steels contain chromium, or chromium and nickel, to render them resistant to corrosion; and basically the amount of chromium controls corrosion resistance (with a minimum of about 12 per cent required) while nickel is added to improve ductility, and increased corrosion resistance is obtained if molybdenum is included. Selection of the correct alloy for the purpose is most important, and advice should be sought in cases of doubt. When corrosion of stainless steel does occur it is usually localised. Resistance to attack by chemicals present in building materials is high (except when mineral acids or chlorides are present) but no stainless steel is completely immune to attack.

1.04 Lead

The corrosion resistance of lead to alkalis is relatively poor and it progressively deteriorates in contact with concrete, particularly in damp conditions. A lead damp proof course laid without protection in concrete may therefore become

* See BRS Digest 109

severely corroded. Lead-cored damp proof courses with a hessian, fibre or asbestos base are, however, satisfactory for general use, provided there is a protective coating (preferably a heavy coat of bitumen).

1.05 Copper

Copper is practically immune from attack, and is not affected to any extent by contact with lime or solutions of calcium hydroxide. It may safely be embedded in fresh concrete, and no destructive action will occur on the area embedded even if the concrete is saturated (provided chlorides are not present).

1.06 Brass

Brass is one of the most important alloys of copper used in building and is primarily an alloy of copper and zinc. There is likely to be a slight attack by plastic concrete which results in the formation of a protective coating on the metal. After the hardening of the concrete, however, further attack is unlikely unless chlorides are present.

1.07 Bronze

Bronze is primarily an alloy of copper and tin although other components may be included. The action of concrete on these alloys is similar to that described for copper and is largely dependent upon their specific composition.

1.08 Aluminium

Aluminium reacts vigorously with wet concrete but when the concrete has set its reactivity is much reduced and aluminium in contact with it is usually not severely attacked. However if the concrete becomes damp then the attack will start again and continue for as long as moist conditions remain. The degree of attack will depend upon the composition of the alloy and the richness of the mix.

Aluminium window frames, copings and sills could therefore be used in contact with concrete without risk of excessive corrosion provided that fresh concrete is prevented from coming into contact with the metal—but it is nevertheless advisable to prevent direct contact either by painting the contact surfaces of the metal with at least two coats of a good quality bitumen paint, or by placing a layer of bitumen felt between the metal and the concrete. Corrosion can lead to cracking of the embedding medium and unsightly salt efflorescence above the level of embedment. Any chlorides used in the mix may accelerate corrosion. Lime mortars affect aluminium in the same way as cement, but less severely. All types of gypsum plaster can cause some corrosion either while they are still wet or if prolonged dampness occurs after the gypsum has dried out.

1.09 Tin

The action of concrete on tin is similar to that for zinc; but this metal is seldom used in direct contact with concrete in the construction industry.

2 Recommendations

When non-ferrous metals are in contact with concrete, the following precautions should be taken (see also table).
● If the concrete is likely to be damp then a protective coating is advisable. If the protective coating is bitumen this will reduce the bond between the metal and the concrete, and should this bond be an important factor then some other means will have to be adopted to ensure secure fixing. Epoxy resins mixed with fine sand may be used as a protective coating to give a satisfactory bond.
● Care should be taken to avoid actual contact or very close proximity between dissimilar metals, including reinforcement, in the concrete as this may lead to galvanic action resulting in corrosion of one of the metals.

Table I Protective measures for metals (from BS CP 3 Chapter IX, 1950).

General corrosive agencies	Metals commonly involved	Measures to be taken
Atmospheric action	Iron and steel	Painting (with or without phosphating) Metallic coatings (zinc, lead, terne, tin, aluminium, etc) used alone or in conjunction with paint Bituminous coatings
	Aluminium	Anodising where conditions are not severe Painting (usually with anodising or chromate treatment where conditions are severe)
Water supplies	Iron and steel	Galvanising Bituminous coatings Treatment of the water
	Lead	Treatment of the water
Soil corrosion (a) General	Iron and steel	Bituminous coatings Concrete casing
(b) Due to sulphate-reducing bacteria in sulphate-bearing soils	Iron, steel or lead	Surrounding with granular filling and draining Bituminous coatings Concrete casing
Juxtaposition or association with other materials (a) Other metals	Galvanised steel Eg galvanised water tanks with copper pipes in the same system, whether electrically connected or not Zinc, galvanised steel or aluminium eg roofs or gutters of these metals which may be attacked by run-off from copper roofs	It is essential that this juxtaposition be avoided
(b) Cement and lime (mortar and concrete)	Lead May be attacked if used in close contact with fresh mortar or concrete when conditions are moist and air is excluded, eg damp-proof courses in walls, water pipes embedded in chases in walls	Metal should be wrapped with paper (to allow penetration of air) or with waterproof felt or paper; or coated with bitumen (to prevent access of alkaline solution to the metal); or packed around with inert material, according to circumstances
	Aluminium or zinc May be attacked by cement or by lime, if embedded or in contact, especially under damp conditions	Access of alkaline solution should be prevented by coating the metal or interposing waterproof felt
(c) Magnesium oxychloride	Iron and steel Aluminium Copper Zinc	Bitumen coating should be applied Seepage of excess gauging liquid to inaccessible steel, or to reinforcement in cracked or porous concrete, should be avoided

Information sheet 18
Repairs to in situ paving

1 General principles

In-situ concrete pavings when properly constructed have a
practically unlimited life. When such pavings do deteriorate,
the indications are that the concrete, the construction or the
workmanship, or possibly all three, were faulty in the first
instance. Any attempt to restore the surface of inferior
concrete by the addition of a thin layer of new material
is likely to result in failure and disappointment.
Preferably repairs should be carried out by adding a
new layer of concrete to the existing paving which will
raise the finished surface. If for any reason the changes in
level which this involves are unacceptable the only solution
is to break up the paving and re-lay it.

2 Procedures

● Carefully examine the existing concrete, first to check
its condition and second to decide to what extent the level
of the surface can be raised. Unless the level of the
existing concrete can be raised by at least 75 mm it will be
necessary to break up the existing concrete, remove the
broken material and prepare the sub-base at the correct
level, and then re-lay the paving.
● Where the level can be raised the required 75 mm
without causing problems, the following work should be
carried out.
1 Thoroughly clean the concrete surface by brushing and
washing with a stiff broom or wire brush. Vegetable matter
such as algae, lichen, moss or weeds adhering to the
surface must be completely removed, and a weedkiller such
as sodium chlorate may be needed in the cracks and joints.
Any particularly smooth parts of the existing surface
should be hacked to help provide a key.
2 The mix for the topping should be 1:2½:4 cement, sand
and coarse aggregate (gravel) with maximum size of 20 mm.
The edges of the new topping will require support until the
concrete has hardened and so formwork must be fixed.
This can be made from 25 mm thick timber, with width
being equal to the depth of the topping. The timbers
should be placed on edge and supported by 450 mm long
pegs driven into the ground at about 1 m centres.
3 The concrete must be spread evenly to a depth of about
12 mm proud of the forms to allow for compaction.
Thorough compaction of the concrete is essential and to
achieve this a tamper should be used. Two passes with the
tamper should give satisfactory compaction. The concrete
should then be level with the tops of the forms, and have a
uniform, close-knit surface. A wood float of the type used
by plasterers is useful for touching up small, rough spots
after compaction.

After compaction is completed, strike off any superfluous concrete by moving forward with a sawing motion from side to side.

4 The joints in the sub-base concrete, and any major transverse cracks, should be carried through the new topping as plain butt joints, but if these are at greater intervals than 3 m then joints should be formed in the new topping at 3 m centres.

5 A suitable finish can be given to the concrete by brushing the surface with a broom immediately after final finishing with the hand tamper (see Information sheet 14).

6 Fresh concrete must not be allowed to dry out too quickly, as this will cause a loss in strength, and possibly surface cracking.

Wet hessian or sacks can be placed directly on to the surface as soon as concreting is completed; and they should be kept wet. Alternatively, polythene sheet or waterproof building paper can be used but the sheets should be well lapped and held down at the edges to prevent draughts. In warm weather, the concrete should be cured for four days.

7 It is not advisable to carry out paving work in very cold weather, that is when the air temperature is at or very close to freezing during the day. The rate of hardening (gain of strength) of concrete is slowed down by low temperatures and if the concrete freezes before it has reached a certain minimum strength it will be permanently damaged.

● Where the level can be raised by the required amount, but the existing concrete is badly cracked or is otherwise in an unsound condition, then the solution is to break up the concrete with a heavy hammer so that the surface resembles a mosaic with 100-150 mm between the cracks in all directions. The broken concrete will then act as a sub-base to the new topping. Every effort should be made at this stage to ensure that a reasonably level sub-base is achieved so that the new concrete to be laid on it will have a uniform thickness.

● When raising the level of concrete paving, care must be exercised to ensure that the final level is not too close to horizontal damp-proof courses in walls; the minimum height between the concrete and the damp-proof course should be 75 mm.

● If it is likely that de-icing salt will be used to remove the snow and ice from the concrete surface, the new topping concrete should be air-entrained.

158

Appendix 1
Site survey notes

The importance of an accurate and comprehensive site survey is that it represents the factual foundation on which the design process will be based. It is therefore essential that a careful inspection and accurate records are made at the outset, specifically related to the client's brief. Obviously superficial observations should be avoided at all costs. Time spent at this stage in assembling records which relate to and supplement each other will be well rewarded during the design process. A thoroughly well organised survey will probably fill out the attached checklists with supplementary information—at worst they may serve only to eliminate interminable internal memos beginning 'Don't forget . . .'. Different methods are used to prepare site notes for survey reports, each varying with inclination, skill or temperament. The best method, and that which is recommended, involves the use of checklists which minimise the possibility of error or omission. Checklists are usually broken down into the sections or elements in which both the report is prepared and the inspection carried out and form not only the basis of inspection but also of the final report. With practice and a set pattern of procedure, the time taken to complete the investigation and obtain all the necessary information can be kept to a minimum without sacrificing thoroughness.

The accompanying checklist sets out the information which, at the very least, should be available. In addition it is essential that the site and surroundings are fully photographed and in particular that the relationships of the site and its environment is well established and readily communicable.

Table: Site Survey: Checklist

Factor	Check	Source (UK)	Record
Title	Full address	client/solicitor/agent	
	History of Site	LPA, MLGD	
	Existing use	LPA	
	Planning or by-laws	LPA	
	Previous consents	LPA	
	Ownership of party walls or fences		
	Restrictive covenants		
	Easements		
	Wayleaves	client/solicitor/agent	
	Rights of way or light	Land registry	
	Rights of air		
	Rights of support		
	Building Regulations & Acts		
Preliminary Survey	Topography	LA, DOE, Aerial	
	Geology	GS or IGS	
	Climatic	MO, Climatological Atlas	
	Soils	SSEW	
	Land capability	MAFF	
	Land Use	LUS	
	Water and Drainage	OS or River Authorities	
	Addresses of adjoining owners	client/solicitor/agent	
	Addresses of public service and Statutory Authorities	LA	
	Future developments	LPA	
	Specific restrictions	LPA, DTI	
	Preservation orders	MLGD	
	Aesthetic control	LPA, RFAC	
	Building or improvement lines	LPA	
	Archaeological and historic	MLGD, Early OS and Nature Conservancy	
Site	Levels and Dimensions		
	Soils		
	Subsidence, Erosion, underground		
	Microclimate	MO	
	Water and Drainage pattern	LA, RA	
	Vegetation (top, middle, lower)		
	Services location and depth	SA	
	Artefacts and local materials	LA and MLGD	
	Materials performance		
	Visual	Isovist records	
	Communications	Transport Authorities	
	Sociological	GRO	
Other Regulations	Commons Act		
	Factory Acts		
	Local Employment Acts		
	National Parks etc Act		
	Offices, Shops & Railway Premises Act		
	Public Health Acts		
Other Authorities	Alkali Inspector		
	British Transport Docks Board		
	British Waterways Board		
	Catchment Area Board		
	Countryside Commission		
	Harbour Board		
	Licensing Authorities		
	Min Ag Fisheries and Food		
	Nature Conservancy		
	Public Health Inspector		
	Public Transport Authorities		
	Water Resources Board		
References	Climatological Atlas of the British Isles	Met Office (HMSO)	
	The investigation of air pollution (annual)	Ministry of Technology (HMSO)	
	Soil survey maps (scale in)	(HMSO)	
	Land classification maps	Min of Ag, Fisheries and Food (HMSO)	
	Land capability maps		
	Land utilisation survey of Great Britain	Geographic Department King's College, London	
	Handbook of Urban Landscape, C. Tandy	The Architectural Press	
Abbreviations	DOE—Department of Environment	MAFF—Ministry of Agriculture, Fisheries and Food	
	DTI—Dept. of Trade and Industry	MLGD—Ministry of Local Government and Development	
	GRO—General Register Office	MO—Meteorological Office	
	GS—Geological Survey	OS—Ordnance Survey	
	IGS—Institute of Geological Science	RA—River Authorities	
	LA—Local Authority	RFAC—Royal Fine Arts Commission	
	LPA—Local Planning Authority	SA—Service Authorities	
	LUS—Land utilisation survey of GB	SSEW—Soils Survey of England and Wales	

Appendix 2 Properties of cements and concretes

Table 1 Cements: properties

Cement type	BS No	Rate of strength development	Rate of heat evolution	Resistance to sulphates
Main types of Portland cement				
Ordinary	12	medium	medium	low
Rapid-hardening	12	high	high	low
Sulphate-resisting	4027	low-medium	low-medium	high-very high
Other types of Portland cement				
Extra-rapid-hardening	—	high—very high	high—very high	low
Ultra high early strength	—	high—very high	high—very high	low
Low heat	1370	low	low	medium-high
Waterproof and water-repellent	—			
Hydrophobic	—	Properties similar to those of ordinary Portland cement		
White	12			
Coloured	12			
Cements containing blastfurnace slag				
Portland-blastfurnace	146	low–medium	low–medium	low–medium
Low heat Portland-blastfurnace	4246	low	low	medium–high
Supersulphated	4248	medium	low	high–very high
High alumina cement	915	very high	very high	very high
Pozzolanic cement	—	low	low–medium	medium–high

Table 2 Properties of various types of concrete using different aggregates

Aggregate	Typical range of dry density Aggregate kg/m³	Concrete kg/m³	Compressive strength N/mm²	Drying shrinkage per cent	Thermal conductivity at 5 per cent moisture content* W/m °C	Main functional requirement
Clinker	700–1050	1050–1500	2–7	0·04–0·08	0·35–0·65	Class 1 fire-resistance / thermal insulation
Exfoliated vermiculite and expanded perlite	60–250	400–1100	0·5–7	0·20–0·35	0·15–0·39	
Pumice	500–900	650–1450	2–15	0·04–0·08	0·21–0·63	
Foamed slag	300–950	950–1500	2–7	0·03–0·07	0·30–0·65	
Expanded clay, shale or slate and sintered pulverised fuel ash	300–1050	700–1300	2–7	0·03–0·07	0·24–0·50	
Foamed slag	500–950	1700–2100	15–60	0·04–0·10	0·85–1·40	strength and durability / Class 2 fire-resistance
Expanded clay, shale or slate and sintered pulverised fuel ash	300–1050	1350–1800	15–60	0·02–0·12	0·55–0·95	
Crushed brick	1100–1350	1700–2150	15–30	—	0·85–1·50	
Crushed limestone	1350–1600	2200–2400	20–80	—	1·6 –2·0	
Flint gravel or crushed stone	1350–1600	2200–2500	20–80	—	1·6 –2·2	

* Calculated from IHVE *Guide*; if available, measured conductivities should be used in preference to these for calculating standard U-values.

Appendix 3 British standards and codes of practice relevant to concrete

Cement
BS 890: 1972 Building limes M
BS 12: — Portland cement (ordinary and rapid-hardening)
Part 1: 1958 Imperial units
Part 2: 1971 Metric units M
BS 146: — Portland-blastfurnace cement
Part 1: 1958 Imperial units
Part 2: 1973 Metric units M
BS 1370: — Low heat Portland cement
Part 1: 1947 Imperial units
Part 2: 1974 Metric units M
BS 4027: 1966 Sulphate-resisting Portland cement
Part 2: 1972 Metric units M
BS 4246: — Low heat Portland-blastfurnace cement
Part 1: 1968 Imperial units
Part 2: 1974 Metric units M
BS 4248: 1974 Supersulphated cement M
BS 4550: — Methods of testing cement
Part 2: 1970 Chemical tests M
BS 4627: 1970 Glossary of terms relating to types of cements, their properties and components

Aggregate
BS 63: — Single-sized roadstone and chippings
Part 1: 1951 Imperial units
Part 2: 1971 Metric units M
BS 812: 1967 Methods for the sampling and testing of mineral aggregates, sands and fillers
BS 877: 1967 Foamed or expanded blastfurnace slag lightweight aggregate for concrete
Part 2: 1973 Metric units M
BS 882, 1201: 1965 Aggregates from natural sources for concrete (including granolithic)
Part 2: 1973 Metric units M
BS 1047: 1952 Air-cooled blastfurnace slag coarse aggregate for concrete
Part 2: 1974 Metric units M
BS 1165: 1966 Clinker aggregate for concrete
BS 1198-1200: 1955 Building sands from natural sources
BS 1438: 1971 Media for biological percolating filters M
BS 1984: 1967 Gravel aggregates for surface treatment (including surface dressings) on roads
BS 2451: 1963 Chilled iron shot and grit
BS 3670: 1963 Methods of sieve analysis of wood-flour M
BS 3681: 1963 Methods for the sampling and testing of lightweight aggregates for concrete
Part 2: 1973 Metric units M
BS 3797: 1964 Lightweight aggregates for concrete
BS 3892: 1965 Pulverised-fuel ash for use in concrete M
BS 4132: 1967 Winkle clinker for landscape work
BS 4619: 1970 Heavy aggregates for concrete and gypsum plaster M

Concrete
BS 1926: 1962 Ready-mixed concrete
BS 1881: — Methods of testing concrete
Part 1: 1970 Methods of sampling fresh concrete M
Part 2: 1970 Methods of testing fresh concrete M
Part 3: 1970 Methods of making and curling test specimens M
Part 4: 1970 Methods of testing concrete for strength M
Part 5: 1970 Methods of testing hardened concrete for other than strength M
Part 6: 1971 Analysis of hardened concrete M
BS 4408: — Non-destructive methods of test for concrete
Part 1: 1969 Electromagnetic cover measuring devices M
Part 2: 1969 Strain gauges for concrete investigations M
Part 3: 1970 Gamma radiography of concrete M
Part 4: 1971 Surface hardness methods M
Part 5: 1974 Measurement of the velocity of ultrasonic pulses in concrete M
BS 4551: 1970 Methods of testing mortars and specification for mortar testing sand M
BS 4721: 1971 Ready-mixed lime: sand for mortar M
PD 6472: 1974 Guide to specifying the quality of building mortars M

Codes of Practice
CP 110 The structural use of concrete
CP 114 Structural use of reinforced concrete in buildings
CP 115 Structural use of prestressed concrete in buildings
CP 2007 Design and construction of reinforced and pre-stressed concrete structures for the storage of water and other aqueous liquids

Glossaries
BS 2787 Glossary for concrete and reinforced concrete
BS 3589 Glossary of general building terms
BS 3683 Glossary of terms used in non-destructive testing.
Parts 1 and 2: 1963.
Part 3: 1964. Parts 4 and 5: 1965
BS 3810 Glossary of terms used in materials handling.
Part 1: 1964. Part 2: 1965. Part 3: 1967. Part 4: 1968
BS 4340 Glossary of formwork terms

Appendix 4 American ASTM standards relevant to concrete

Cement
Specifications for
*D 545 67 (1972) Expansion Joint Fillers, Preformed, for Concrete (Non-extruding and Resilient Types)
*C 156 74 Water Retention by Concrete Curing Materials
*C 595 74 Blended Hydraulic Cements
*C 230 68 (1974) Flow Table for Use in Texts of Hydraulic cement
*C 10 73 Natural Cement
C 150 74 Portland Cement

Aggregates
Specifications for
*C 33 74 Concrete Aggregates
D 2940 74 Graded Aggregate Material for Bases or Subbases for Highways or Airports
*C 35 70 Gypsum Plaster, Inorganic Aggregates for Use in
*C 331 69 Lightweight Aggregates for Concrete Masonry Units
*C 332 66 (1971) Lightweight Aggregates for Insulating Concrete
*C 330 69 Lightweight Aggregates for Structural Concrete
C 404 70 Masonry Grout, Aggregates for
C 144 70 Masonry Mortar, Aggregate for
C 637 73 Radiation-Shielding Concrete, Aggregates for
*D 1241 68 Soil-Aggregate Subbase, Base, and Surface Courses, Materials for
D 448 54 (1973) Standard Sizes of Coarse Aggregate for Highway Construction
*E 11 70 Wire-Cloth Sieves for Testing Purposes

Methods of Test for
*C 535 69 Abrasion of Large Size Coarse Aggregate by Use of the Los Angeles Machine, Resistance to
*C 131 69 Abrasion of Small Size Coarse Aggregate by Use of the Los Angeles Machine, Resistance to
*C 227 71 Alkali Reactivity, Potential, of Cement-Aggregate Combinations (Mortar-Bar Method)
*D 1411 69 Chlorides, Water-Soluble, Present as Admixes in Graded Aggregate Road Mixes
*C 142 71 Clay Lumps and Friable Particles in Aggregates
D 3042 72 Insoluble Residue in Carbonate Aggregates
*C 123 69 Lightweight Pieces in Aggregate
*C 117 69 Materials Finer Than No 200 (75μm) Sieve in Mineral Aggregates by Washing
C 566 67 (1972) Moisture Content, Total, of Aggregate by Drying
*C 87 69 Organic Impurities in Fine Aggregate on Strength of Mortar, Effect of
*C 40 73 Organic Impurities in Sands for Concrete
C 586 69 Potential Alkali Reactivity of Carbonate Rocks for Concrete Aggregates (Rock Cylinder Method)
*C 289 71 Potential Reactivity of Aggregates (Chemical Method)
C 702 71 T Reducing Field Samples of Aggregate to Testing Size
*D 75 71 Sampling Aggregates
C 235 68 Scratch Hardness of Coarse Aggregate Particles
*C 136 71 Sieve or Screen Analysis of Fine and Coarse Aggregates
*D 451 63 (1970) Sieve Analysis of Granular Mineral Surfacing for Asphalt Roofing and Shingles
*D 452 63 (1970) Sieve Analysis of Nongranular Mineral Surfacing for Asphalt Roofing and Shingles

Methods of Test for
*C 88 73 Soundness of Aggregates by Use of Sodium Sulfate or Magnesium Sulfate
*C 127 73 Specific Gravity and Absorption of Coarse Aggregate
*C 128 73 Specific Gravity and Absorption of Fine Aggregate
C 641 71 Staining Materials in Lightweight Concrete Aggregates
*C 70 73 Surface Moisture in Fine Aggregate
*C 29 71 Unit Weight of Aggregate
*C 30 37 (1970) Voids in Aggregate for Concrete
*C 342 67 (1973) Volume Change, Potential, of Cement-Aggregate Combinations

Recommended Practices for
C 682 71 T Evaluation of Frost Resistance of Coarse Aggregates in Air-Entrained Concrete by Critical Dilation Procedures
*C 295 65 (1973) Petrographic Examinations of Aggregates for Concrete

---

* Approved as **American National Standard** by American National Standards Institute.

Definitions of Terms Relating to:
E 12 70 Density and Specific Gravity of Solids, Liquids, and Gases
*C 125 74 Concrete and Concrete Aggregates
C 638 73 Constituents of Aggregates for Radiation-Shielding Concrete, Descriptive Nomenclature of
*C 294 69 Constituents of Natural Mineral Aggregates, Descriptive Nomenclature of
*D 8 73 Materials for Roads and Pavements

Concrete
Specifications for:
*C 260 73 Air-Entraining Admixtures for Concrete
*C 490 74 Apparatus for Use in Measurement of Length Change of Hardened Cement Paste, Mortar, and Concrete
C 494 71 Chemical Admixtures for Concrete
C 685 74 T Concrete Made by Volumetric Batching and Continuous Mixing
*C 387 69 Dry Combined Materials, Packaged, for Mortar and Concrete
*C 618 73 Fly Ash and Raw or Calcined Natural Pozzolans for Use in Portland Cement Concrete
C 511 73 Moist Cabinets and Rooms Used in the Testing of Hydraulic Cements and Concretes
*C 94 74 Ready-Mixed Concrete
*C 470 73 T Molds for Forming Concrete Test Cylinders Vertically

Methods of test for:
*C 418 68 (1974) Abrasion Resistance of Concrete
C 779 74 Abrasion Resistance of Horizontal Concrete Surfaces
*C 138 74 Air Content (Gravimetric), Unit Weight, and Yield of Concrete
*C 173 73a Air Content of Freshly Mixed Concrete by the Volumetric Method
*C 231 73 Air Content of Freshly Mixed Concrete by the Pressure Method
*C 233 73 Air-Entraining Admixtures for Concrete
*C 227 71 Alkali Reactivity, Potential, of Cement-Aggregate Combinations (Mortar-Bar Method)
*C 360 63 (1968) Ball Penetration in Fresh Portland Cement Concrete
*C 232 71 Bleeding of Concrete
C 234 71 Bond Developed with Reinforcing Steel, Comparing Concretes on the Basis of
C 617 73 Capping Cylindrical Concrete Specimens
C 683 74 T Compressive and Flexural Strength of Concrete Under Field Conditions
*C 116 (1974) Compressive Strength of Concrete Using Portions of Beams Broken in Flexure
C 513 69 Compressive Strength, Hardened Lightweight Insulating Concrete for, Securing, Preparing, and Testing Specimens from
C 495 69 Compressive Strength of Lightweight Insulating Concrete
*C 39 72 Compressive Strength of Cylindrical Concrete Specimens
*C 31 69 Concrete Test Specimens, Making and Curing in the Field
*C 192 69 Concrete Test Specimens, Making and Curing in the Laboratory
*C 512 74 Creep of Concrete in Compression
C 671 72 T Critical Dilation of Concrete Specimens Subjected to Freezing
*C 42 68 (1974) Drilled Cores and Sawed Beams of Concrete, Obtaining and Testing
*C 78 64 (1972) Flexural Strength of Concrete (Using Simple Beam with Third-Point Loading)
*C 293 68 (1974) Flexural Strength of Concrete (Using Simple Beam with Center-Point Loading)
*C 311 68 (1974) Fly Ash for Use as an Admixture in Portland Cement Concrete, Sampling and Testing
*C 215 60 (1970) Fundamental Transverse, Longitudinal, and Torsional Frequencies of Concrete Specimens
*C 85 66 (1973) Hardened Portland Cement Concrete, Cement Content of
C 157 74 Length Change of Hardened Cement Mortar and Concrete
*C 174 49 (1969) Length of Drilled Concrete Cores, Measuring
C 684 74 Making, Accelerated Curing, and Testing of Concrete Compression Test Specimens
C 441 69 Mineral Admixtures, Effectiveness of, in Preventing Excessive Expansion of Concrete Due to Alkali-Aggregate Reaction

C 597 71 Pulse Velocity Through Concrete

C 666 73 Resistance of Concrete to Rapid Freezing and Thawing

*C 172 71 Sampling Fresh Concrete

C 672 72 T Scaling Resistance of Concrete Surfaces Exposed to Deicing Chemicals

*C 143 71 Slump of Portland Cement Concrete

C 642 69 T Specific Gravity, Absorption, and Voids in Hardened Concrete

*C 496 71 Splitting Tensile Strength of Cylindrical Concrete Specimens

*C 469 65 (1970) Static Modulus of Elasticity and Poisson's Ratio of Concrete in Compression

*C 403 70 Time of Setting of Concrete Mixtures by Penetration Resistance

C 567 71 Unit Weight of Structural Lightweight Concrete

*E 4 72 Verification of Testing Machines

C 341 74 Length Change of Drilled or Sawed Specimens of Cement Mortar and Concrete

C 342 67 (1973) Volume Change, Potential, of Cement-Aggregate Combinations

C 138 74 Unit Weight, Yield, and Air Content (Gravimetric) of Concrete

Recommended Practices for:

C 457 71 Air-Void Content, and Parameters of the Air-Void System in Hardened Concrete, Microscopical Determination of

C 670 71 T Preparing Precision Statements for Test Methods for Construction Materials

*E 329 72 Inspection and Testing Agencies for Concrete, Steel, and Bituminous Materials as Used in Construction

C 682 71 T Evaluation of Frost Resistance of Coarse Aggregates in Air-Entrained Concrete by Critical Dilation Procedures

Definitions of Terms Relating to:

*C 125 74 Concrete and Concrete Aggregates

Appendix 5
Initial enquiry for the
supply of ready mixed concrete

Information to be provided by the purchaser for all enquiries

1	**CONTRACTOR**										
2	**NAME AND LOCATION OF CONTRACT**										
3	**TYPE OF CONTRACT**										
4	**SPECIFYING AUTHORITY**										
5	**STARTING DATE (approx)**										
6	**ESTIMATED CONSTRUCTION PERIOD**										
7	**QUANTITIES PER DAY m³ (approx)**	Average: Maximum:									
8	**PRODUCTION**										
9	**CONCRETE MIX**										
	Designed mix **Grade** (Strength at 28 days (N/mm²))										
or	**Compliance rules** (ie Standard or Code) **Prescribed mix** **Grade** (Cement content, kg/m³)										
	Compliance rules	Cement content stated on delivery ticket									
10	**WORKABILITY** (Slump mm)										
11	**MATERIALS**										
	Type of cement										
	Nominal max. agg. size (mm)										
12	**CONCRETE QUANTITIES** (m³ for each mix)										

Appendix 6
Metric conversion factors

Basic units

Length	1 yd	=	0·9144 m
	1 m	=	1·0936 yd
Mass	1 lb⎤	=	0·45359 kg
	1 kg	=	2·2046 lb
Capacity	1 pt	=	0·56825 litres
	1 litre	=	1·76 pt
Temperature	1 deg F	=	0·5556 deg C
	1 deg C	=	1·8000 deg F
Force	1 lbf	=	4·4482 N
	1 N	=	0·22481 lbf
	1 lbf	=	0·5359 kgf
	1 kgf	=	2·2046 lbf

Additional familiar units

Length	1 mile	=	1·6093 km
	1 km	=	0·6214 mile
	1 ft	=	0·3048 m
	1 m	=	3·2808 ft
	1 in	=	25·4 mm
	1 mm	=	0·03937 in
Mass	1 ton	=	1016·05 kg
		=	1·01605 tonne
	1 tonne	=	
	1000 kg	=	0·9842 ton
	1 cwt	=	50·802 kg
	1 kg	=	0·01968 cwt
	1 oz	=	28·3495 g
	1 g	=	0·03527 oz
Capacity	1 gal	=	4·54596 litres
	1 litre	=	0·220 gal
Area	1 mile²	=	2·58999 km²
	1 km²	=	0·3861 mile²
	1 yd²	=	0·83613 m²
	1 m²	=	1·19599 yd²
	1 ft²	=	0·09290 m²
	1 m²	=	10·7639 ft²
	1 in²	=	645·16 mm²
	1 mm²	=	0·00155 in²
Volume	1 yd³	=	0·76455 m³
	1 m³	=	1·30795 yd³
	1 ft³	=	0·02832 m³
	1 m³	=	35·315 ft³
	1 in³	=	16387·1 mm³
	1 mm³	=	0·0000610 in³

Derived units

Velocity	1 mph	=	1·6093 km/h
	1 km/h	=	0·6214 mph
	1 ft/sec	=	0·3048 m/sec
	1 m/sec	=	3·2808 ft/sec
Mass per Unit Length	1 lb/ft	=	1·4882 kg/m
	1 kg/m	=	0·672 lb/ft
Mass per Unit Area	1 lb/ft²	=	4·8824 kg/m²
	1 kg/m²	=	0·2048 lb/ft²
Mass per Unit Volume (Density)	1 lb/yd³	=	0·5933 kg/m³
	1 kg/m³	=	1·6856 lb/yd³
	1 lb/ft³	=	16·0185 kg/m³
	1 kg/m³	=	0·06243 lb/ft³
	1 lb/in³	=	0·02768 g/mm³
	1 g/mm³	=	36·13 lb/in³
Fuel Consumption	1 mpg	=	0·3540 km/l
	1 km/l	=	2·825 mpg
Stress	1 lbf/in²	=	0·006895 N/mm²
	1 N/mm²	=	145·038 lbf/in²
	1 lbf/in²	=	0·000703 kgf/mm²
	1 kgf/mm²	=	1422·3 lbf/in²

Capacity per Mass	1 pt/112 lb	=	0·5593 litres/50 kg
	1 litre/50 kg	=	1·7880 pt/112 lb
	1 fl oz/112 lb	=	27·96 ml/50 kg
	1 ml/50 kg	=	0·0358 fl oz/112 lb
Capacity per Unit Volume	1 gal/yd³	=	5·9459 litres/m³
	1 litre/m³	=	0·16818 gal/yd³
	1 pt/yd³	=	0·74324 litres/m³
	1 litre/m³	=	1·34551 pt/yd³
Price	1 £p/yd³	=	1·30795 £p/m³
	1 £p/m³	=	0·76455 £p/yd³

References

1 Cullen, Gordon. *The Concise Townscape*, London, 1971, Architectural Press. £1·95
2 Tandy, Clifford (editor). *Handbook of Urban Landscape*, London, 1972, Architectural Press, £7·50.
3 Fairbrother, Nan. *New Lives, New Landscapes*, London, 1970, Architectural Press, £4·50; also *The Nature of Landscape Design* 1974, £5·75.
4 Handbook of Building Environment, *The Architects' Journal* 2.10.1968 to 13.8.1969.
5 Parkin, P. H. and Humphreys, H. R. *Acoustics, Noise and Buildings*, London, 1969, Faber and Faber (paperback).
6 Elder, A. J. and Vandenberg, Maritz (editors) *AJ Handbook of Building Enclosure*, London, 1974, Architectural Press, £10·50 (paperback £5·95).
7 Newman, Oscar. *Defensible Space*, London, 1973, Architectural Press, £5·75.
8 Beazley, Elisabeth. *Design and Detail of the Space between Buildings*. London, 1960 (3rd impression 1968), Architectural Press, o.p.; available on microfiche at £4·00.
9 Fairweather, Leslie and Sliwa, Jan. *AJ Metric Handbook*. Third Edition. London, 1971, Architectural Press, £2·75.
10 Hodgkinson, Allan (editor). *AJ Handbook of Building Structure*. London, 1974, Architectural Press, £4·50 (paperback).
11 Gage, Michael. *Guide to Exposed Concrete Finishes*. London, 1970, Architectural Press, £3·75.
12 Gage, Michael, and Kirkbride, Tom. *Design in Blockwork*. London, Architectural Press (2nd edition 1976) approx £4·50.

Photo credits

9 p, s, C & CA
10 p, s, C & CA
12 a, A. L. Hobson, p, S. Hunt, s, C & CA
13 p, Bill Duxbury, s, C & CA
15 (a) p, s, C & CA
15 (b) p, S. W. Newbery, s, C & CA
15 (c) s, C & CA
16 (a) p, s, C & CA
16 (b) p, s, C & CA
16 (c) p, s, C & CA
17 (a) p, Wexham Springs, s, C & CA
17 (b) s, C & CA
Information sheet 4
1 (a) p, s, C & CA
1 (b) p, s, C & CA
2 p, s, C & CA
3 p, s, C & CA
4 p, s, C & CA
Information sheet 5
1 (a) p, s, C & CA
1 (b) p, s, C & CA
1 (c) p, s, C & CA
1 (d) p, s, C & CA
1 (e) p, s, C & CA
3 p, s, C & CA
Information sheet 6
1 (a) p, s, C & CA
1 (b) a, p, Mono Concrete
4 a, British Steel Corporation
5 a, Wexham, p, s, C & CA
6 a, p, Mono Concrete
10 a, Glascrete Pavement Lights
Information sheet 7
1 (a) p, B. Duxbury, s, C & CA
1 (b) p, s, C & CA
4 (a) p, S. W. Newbery, s, C & CA
4 (b) a, R. Seifert & Partners
Information sheet 9
1 (a) p, B. Duxbury, s, C & CA

1 (b) p, A. Court
2 p, A. Court
Information sheet 10
1 s, C & CA, p, Crispin Eurich
Information sheet 11
1 p, s, C & CA
2 s, Hodge Clemco Ltd
Information sheet 12
1 s, C & CA
2 p, Bill Duxbury, s, C & CA
3 s, C & CA
4 s, Errut Ltd
5 s, Tremix
6 p, Alexandra Studio, s, Errut Ltd
Information sheet 13
1 p, Bill Duxbury, s, C & CA
2 p, Bill Duxbury, s, C & CA
4 p, Bill Duxbury, s, C & CA
5 p, Bill Duxbury, s, C & CA
Information sheet 14
1 a, O.P.C. & Hestor gravel, p, s, C & CA
2 p, Bill Duxbury, s, C & CA
3 s, Gee Walker & Slater Ltd
4 a, s, William Mitchell Design Consultants
5 p, s, C & CA
6 a, Wexham Springs
7 p, S. W. Newbery, s, C & CA
9 p, Greville Photography
10 p, Birmingham Post Studios
11 p, Leadbetter, s, Errut Ltd
Information sheet 15
1 s, C & CA
2 a, s, William Mitchell Design Consultants
Information sheet 16
1 p, s, C & CA
2 p, s, C & CA
3 p, s, C & CA

DATE DUE

DEMCO NO. 38-298